New Studies in Multimodality

Also available from Bloomsbury:

Introduction to Multimodal Analysis, David Machin
Multimodal Discourse, Gunther Kress and Theo van Leeuwen
Multimodal Teaching and Learning, Gunther Kress, Carey Jewitt, Jon Ogborn and Tsatsarelis Charalampos

New Studies in Multimodality

Conceptual and Methodological Elaborations

Edited by Ognyan Seizov and Janina Wildfeuer

Bloomsbury Academic
An imprint of Bloomsbury Publishing Plc

B L O O M S B U R Y
LONDON • OXFORD • NEW YORK • NEW DELHI • SYDNEY

Bloomsbury Academic
An imprint of Bloomsbury Publishing Plc

50 Bedford Square
London
WC1B 3DP
UK

1385 Broadway
New York
NY 10018
USA

www.bloomsbury.com

BLOOMSBURY and the Diana logo are trademarks of Bloomsbury Publishing Plc

First published 2017

British Library Cataloguing-in-Publication Data
A catalogue record for this book is available from the British Library.

ISBN: HB: 978-1-3500-2651-3
ePDF: 978-1-3500-2653-7
ePub: 978-1-3500-2652-0

Library of Congress Cataloging-in-Publication Data
Names: Seizov, Ognyan, editor. | Wildfeuer, Janina, 1984- editor.
Title: New studies in multimodality : conceptual and methodological elaborations / edited by Ognyan Seizov and Janina Wildfeuer.
Description: London ; New York : Bloomsbury Academic, [2017] | Includes bibliographical references and index.
Identifiers: LCCN 2017009500| ISBN 9781350026513 (hb) | ISBN 9781350026537 (epdf) | ISBN 9781350026520 (epub)
Subjects: LCSH: Modality (Linguistics) | Communication–Methodology. | Communication--Psychological aspects. | Language and languages--Study and teaching. | Semiotics--Social aspects.
Classification: LCC P99.4.M6 N39 2017 | DDC 302.2--dc23 LC record available at https://lccn.loc.gov/2017009500

Typeset by Deanta Global Publishing Services, Chennai, India

Table of Contents

List of Figures

List of Tables

Notes on Contributors

Morten Boeriis, PhD in multimodality and moving images, is an associate professor at the University of Southern Denmark, Denmark, in the areas of visual communication, multimodality, and business communication. He teaches a variety of courses on visual communication and analysis at Business Communication Studies and Film and Media Studies at University of Southern Denmark. He has edited the Routledge book *Social Semiotics – Key Figures, New Directions* (2015) (with Andersen, Maagerø, and Tønnessen), the book *Nordisk Socialsemiotik – pædagogiske, multimodale og sprogvidenskabelige landvindinger* (2012) (with Andersen) on Syddansk Universitetsforlag, and the Taylor & Francis book *Advancing Multimodal and Critical Discourse Studies* (2017) (with Zhao, Björgvall, and Djonov). He has written articles and presented papers around the world on topics concerning audiovisual social semiotics and multimodality. He has been working in TV production and as a freelance photographer.

Hans-Jürgen Bucher, PhD in linguistics, is a professor of media studies at the University of Trier, Germany, in the areas of audience research, media, and multimodal discourse analysis, political communication, and science communication. He has worked as a journalist for newspapers and radio stations and as a teacher at a journalist school. Besides researching on reception and audience studies of print, online, and audiovisual media, he has conducted research projects in science communication ("interactive science"), network communication in social media, on the internet in China, and the quality of television programs. His publications include books on interactional reception theory, media formats, and media history, as well as journal articles and book sections on multimodal discourse analysis, reception studies, political communication, journalism, and visual media.

Marc Debus, Dr. rer. soc., is a professor of comparative government at the University of Mannheim, Germany. His research interests include political institutions, in particular multilevel systems, and their effects on the political behavior of voters and legislators, as well as party competition, coalition politics, and decision-making within parliaments and governments. His recent

List of Tables

Notes on Contributors

Morten Boeriis, PhD in multimodality and moving images, is an associate professor at the University of Southern Denmark, Denmark, in the areas of visual communication, multimodality, and business communication. He teaches a variety of courses on visual communication and analysis at Business Communication Studies and Film and Media Studies at University of Southern Denmark. He has edited the Routledge book *Social Semiotics – Key Figures, New Directions* (2015) (with Andersen, Maagerø, and Tønnessen), the book *Nordisk Socialsemiotik – pædagogiske, multimodale og sprogvidenskabelige landvindinger* (2012) (with Andersen) on Syddansk Universitetsforlag, and the Taylor & Francis book *Advancing Multimodal and Critical Discourse Studies* (2017) (with Zhao, Björgvall, and Djonov). He has written articles and presented papers around the world on topics concerning audiovisual social semiotics and multimodality. He has been working in TV production and as a freelance photographer.

Hans-Jürgen Bucher, PhD in linguistics, is a professor of media studies at the University of Trier, Germany, in the areas of audience research, media, and multimodal discourse analysis, political communication, and science communication. He has worked as a journalist for newspapers and radio stations and as a teacher at a journalist school. Besides researching on reception and audience studies of print, online, and audiovisual media, he has conducted research projects in science communication ("interactive science"), network communication in social media, on the internet in China, and the quality of television programs. His publications include books on interactional reception theory, media formats, and media history, as well as journal articles and book sections on multimodal discourse analysis, reception studies, political communication, journalism, and visual media.

Marc Debus, Dr. rer. soc., is a professor of comparative government at the University of Mannheim, Germany. His research interests include political institutions, in particular multilevel systems, and their effects on the political behavior of voters and legislators, as well as party competition, coalition politics, and decision-making within parliaments and governments. His recent publications have appeared, among other journals, in *The Journal of Politics*, *Public Choice*, *Political Science Research and Methods*, *West European Politics*, *Party Politics*, *Electoral Studies*, and *Legislative Studies Quarterly*.

Victor Lim-Fei, PhD, has been working in the field of multimodal discourse analysis and linguistics for over a decade. He is presently deputy director and lead specialist for educational technology, Ministry of Education, Singapore, where he has experience in translational research, policy formulation, and program development, with a focus on how educational technology can improve learning in the classroom. Victor is passionate about education and its value in improving both personal and societal well-being. Victor's publications include invited book chapters, articles, and journal papers. He is also a peer reviewer of books and journals in the field of multimodal research.

Arne Krause, M.A. in linguistics, is a researcher at Hamburg University, Germany, in the areas of functional pragmatics, scientific communication, institutional communication, and multimodality. He conducts classes on German linguistics and institutional communication. His publications include studies on academic teaching and learning at German and Italian universities and the analysis of visual media in academic teaching; he is presently pursuing a PhD on this topic with a transdisciplinary approach.

Carmen Daniela Maier, PhD, is an associate professor, and a member of Center of Corporate Communication at the Department of Business Communication, School of Business and Social Sciences, Aarhus University, Denmark. Among her latest publications are the chapters "A multimodal analysis of the environment beat" in *Critical Multimodal Studies* (Routledge, 2013) and "Stretching the multimodal boundaries of professional communication" in *The Routledge Handbook of Language and Professional Communication* (Routledge, 2014). She is the coeditor of *Interactions, Images and Texts: A Reader in Multimodality* (De Gruyter, 2014). She has also edited the thematic issue *Multimodality, synesthesia and intersemiotic translation* of *Hermes: Journal of Language and Communication in Business* (2016). Her research areas include corporate communication, knowledge communication, environmental communication, and multimodality.

Yannik Porsché, PhD in sociology, does research at the Goethe University Frankfurt/Main and the Humboldt University of Berlin on social, cultural, and organizational forms of knowledge generation and circulation in criminal prevention work of the police. His PhD at the University of Mainz and the Université de Bourgogne in Dijon dealt with museum exhibitions about public representations of immigrants in France and Germany. He combines methods of interaction analysis, discourse analysis, and ethnography in the multimodal analysis of social interactions in institutional contexts. His publications include a monograph entitled *Public Representations of Immigrants in Museums - Exhibition and Exposure in France and Germany* (Palgrave, 2017) and a collaborative book entitled *Polizeilicher Kommunitarismus. Eine Praxisforschung urbaner Kriminalprävention* (Campus, 2016).

Søren Vigild Poulsen, PhD in multimodal web analysis, is an assistant professor at the University of Southern Denmark, Denmark; His main area of focus is multimodality. He teaches courses on semiotics, digital and social media marketing, web design, digital communication, rhetoric, crises communication, and communication planning. He has written a monograph on social semiotic and cognitive analysis of websites as multimodal texts (2014) as well as articles and papers on digital media. He is currently working on a project about social media as semiotic technology.

Ognyan Seizov, PhD in communication science, is a research associate at the Contract Management Institute, SRH Hochschule Berlin, Germany. His academic work spans political communication online, multimodal document analysis, media reception, and mixed-method research. He is the author of *Political Communication Online: Structures, Functions, and Challenges* (Routledge, 2014), and his most recent project tackled the image-text relations and various persuasion strategies found in US, UK, and German political blogs.

Angelika Storrer is full professor and head of the Department of German Linguistics at the University of Mannheim, Germany. Her research interests cover corpus linguistics, e-lexicography, and linguistic aspects of computer-mediated communication and social media. She is active in various interdisciplinary research projects and networks on digital humanities and corpus linguistics, for example, as a member of the advisory board of DARIAH-DE (Digital Infrastructure for the Arts and Humanities—German section) and as a participant in working groups of CLARIN-D (Common Language Resources and Technology Infrastructure—German section). She is a member of the

Berlin-Brandenburg Academy of Sciences and Humanities (BBAW), Germany, where she is involved in corpus and dictionary projects.

Heiner Stuckenschmidt, Dr. rer.nat., is full professor for computer science at the University of Mannheim, Germany, where he heads the Artificial Intelligence research group. His group is conducting fundamental and applied research in knowledge representation and knowledge management. He is author of the first German textbook on ontologies and has presented and published more than 100 papers in international peer-reviewed conferences and journals. In the last couple of years, he has been involved in a number of interdisciplinary research projects in cooperation with researchers from political science, economics, media and communication studies, and linguistics. He is part of the Data and Web Science Group at the University of Mannheim and coeditor in chief of Springer's Journal on Data Semantics. His current interests include the use of background knowledge for data analysis and integration as well as the application of natural language processing for supporting the social sciences.

Serene Tan Kok Yin, MA in English, is interested in innovation in English language and literature teaching. Presently with the Ministry of Education, Singapore, Serene works closely with educators to explore innovative approaches to teaching English language and literature, with a focus on leveraging information and communications technology to enhance learning. Apart from championing the importance of multimodal literacy in education, she is working with schools to integrate multimodality into existing English curricula. Other interests include postcolonial and subaltern studies, South African literature, and theorizations of literature pedagogy.

Christopher Taylor is full professor of English Language and Translation in the Department of Law and Languages at the University of Trieste, Italy. He is a deputy director of the University Language Centre in Trieste and was the president of the national association (AICLU) from 2007 to 2010. He has a long and remarkable track record in the field of translation for many years as his many articles and books—for example, *Language to Language* (1998, Cambridge University Press)—on the subject demonstrate. Film translation, in its many aspects, has been his major pursuit in recent years with significant publications relating to such issues as dubbing, subtitling, and localization, and more recently audiovisual translation for the aurally challenged and audio description for the visually challenged. His numerous publications in this field include several papers on multimodal transcriptions and translations. In 2012 he published with Elisa Perego the volume *Tradurre l'audiovisivo*. He has both participated in and organized numerous international conferences, including *Tradurre il Cinema* in Trieste, the *Convegno Nazionale AICLU* in Trieste, and the *European Systemic Functional Linguistics Conference & Workshop* in Gorizia, both in Italy. He has also been a national coordinator of the Italian research projects *Linguatel* and *Didactas* and has recently coordinated a European Union project Audio Description: Lifelong Access for the Blind (ADLAB), which achieved 'Success Story' status.

Christoph Kilian Theil holds a bachelor of arts in culture and economy from the University of Mannheim, Germany. For his bachelor thesis, "Finance and Fiction—About the Self-Referentiality of Money Based on Goethe's 'Faust II' and Economic Theory," he earned the highest endowed award for theses in humanities at his university. He is currently pursuing a master of science in business administration and is working as a research assistant at the Chair of Artificial Intelligence in the Data and Web Science Group of the University of Mannheim. In his interdisciplinary research, he applies natural language processing methods to corpora in political and business contexts. His work involves the sentiment analysis of Wikipedia articles and analysis of changelogs of mobile applications as well as the automatic detection of uncertainty in earnings calls.

Theo van Leeuwen is an emeritus professor at the University of Technology Sydney, Australia, a professor of language and communication at the University of Southern Denmark, Denmark, and a honorary professor at the University of Lancaster, England, and the University of New South Wales, Australia. He has published widely on critical discourse analysis, multimodality, and social and visual semiotics. His books include *Reading Images—The Grammar of Visual Design* (with Gunther Kress; 1996); *Multimodal Discourse* (with Gunther Kress; 2001); *Introducing Social Semiotics* (2004); *Speech, Music, Sound* (1999); *Global Media Discourse* (with David Machin; 2007); *The Language of Colour* (2011); and *Discourse and Practice* (2008). He is a founding editor of the journal *Visual Communication.*

Hartmut Wessler is a professor of media and communication studies at the University of Mannheim, Germany, and a member of the Mannheim Center for European Social Research (MZES). His research focuses on the comparative analysis of political communication, transnational communication, and multimodal media analysis. He uses human coding as well as automated approaches to analyzing both textual and visual materials. Hartmut Wessler is the author of *Öffentlichkeit als Prozess [The Public Sphere in Processual Perspective]* (Westdeutscher Verlag, 1999), and coauthor of *Transnationalization of Public Spheres* (Palgrave Macmillan, 2008) and *Transnationale Kommunikation* (Springer VS, 2012). He edited and coedited eight books, including the *International Encyclopedia of Political Communication* (Wiley Blackwell, 2015), and wrote numerous book chapters and articles, which have, among others, appeared in *Journal of Communication, Communication Theory, Political Communication,* and *The International Journal of Press/Politics.*

Janina Wildfeuer, PhD in linguistics, is a researcher at Bremen University, Germany, in the areas of multimodality and digital media studies as well as linguistics, discourse analysis, and semiotics. She teaches classes in multimodal, interdisciplinary, and applied linguistics and analyzes films, comics, online discourses, and other multimodal documents within several projects exploring the notion of multimodal discourse. Her publications include a monograph, *Film Discourse Interpretation* (2014, Routledge), the edited volumes *Building Bridges for Multimodal Research* (2015, Peter Lang), *Film Text Analysis* (2017, Routledge), *Mapping Multimodal Performance Studies* (2017, Routledge) as well as several contributions and papers, mostly focusing on interdisciplinary approaches in the multimodality context and bridge-building aspects in the humanities and beyond.

Preface

We came to do this book on multimodal studies, with its special focus on readdressing concepts, debates, and methods from a number of diverse fields, in large part thanks to the multidisciplinarity, diversity, and bridge-building that characterize our own academic careers. We entered multimodal territory from very different angles (linguistics and visual communication); we got acquainted with many of its tools and adapted them to our topics of interest (among others: films, comics, political communication online, and digital persuasion); and to this day we keep discovering new synergy potentials among new fields and growing numbers of driven colleagues as we watch multimodality expand. We aimed to put modern multimodal studies' inherent openness and our awe of its diversity and breadth at the center of this rich volume, and we hope the wealth of ideas and approaches will please the reader as much as it did us.

This book contains a selection of papers originally presented at the Second Bremen Conference on Multimodality (BreMM15), which we hosted at the University of Bremen, Germany, in September 2015. This event was part of a conference series that started in 2014 and continues with BreMM17, planned for September 2017. We take this high level of continuity and commitment seriously, and we believe it is one more example of how much playing room multimodality creates in today's communication study environment.

As any other academic product, this volume would not be here without some additional help and support. We would like to express our thanks to all authors for delivering their fascinating, mind-expanding talks and later turning them into the chapters you are about to read; the anonymous reviewers who have provided us with apt feedback on our initial book proposal and on the content of the individual chapters; the Bremen Institute for Transmedial Textuality Research (BITT) for its financial support of the BreMM15 conference; and the University of Bremen for continually providing us with an academic home base.

See you at BreMM17 then!

Ognyan Seizov and Janina Wildfeuer, the editors,
Bremen, Germany, March 2017

1

Introduction: Rethinking Multimodality in the Twenty-first Century

Ognyan Seizov and Janina Wildfeuer

1 Introduction

Even if it is still young, the twenty-first century has brought about sweeping changes to human communication in all its forms, and science and research have been quick to follow with a range of developments for characterizing, quantifying, classifying, and explaining the fast-paced evolution we are witnessing, especially in mediated communication. At the same time, new media and new genres have pushed us to reassess our evaluations of, and approaches to, older media as well, as McLuhan (1960) has long postulated. Few other fields of research are better positioned than *multimodality* to broaden our understanding and mastery of evolving communication forms, media, and genres. With an inclination to large-N, corpus-based studies, predisposition to partial or full automation, hardwired attention to detail, acute awareness of communication's social embedding, and theories of meaning making that are not medium- or genre-specific, the area of multimodal studies today is excellently equipped to tackle the changes across the communication canvas of modern time.

There are, however, a few caveats that beg the question whether the account above is not overly optimistic. For one, the media developments we are witnessing have a markedly global character, in large part thanks to the World Wide Web's overarching influence both as a medium and an infrastructure, upon which new media, genres, and channels are built. In contrast, multimodality often remains more or less wrapped in localized theories and regional schools of thought, variously pitting the Anglo-Saxon against the German approach to multimodal studies (cf. Wildfeuer 2015), or the traditional semiotic against the pragmatic-discursive view of human communication (cf. Bucher in this volume), to name but two prominent divides. Debates abound when it comes to defining core terms such as "mode," "meaning," "sign," or even "multimodality" itself. In order to rise to the occasion and realize its undoubted potential, therefore, multimodality needs a critical mass of practitioners who reach across disciplinary and regional demarcations in order to bring their perspectives closer together and open their sometimes narrow definitions and views to tried and tested practices, which come from the other side. The ultimate goal of this openness is the amalgamation of basic principles and methods into a discipline of multimodality that is fit to continue delivering insights and scientific explanations of old, new, and future media practices on a global scale.

Addressing the divisions and caveats above is based in interdisciplinary and international dialogue. The landscape of international conferences and events in the field is already adapting to this necessity, offering a number of events that cross disciplinary lines with the intention to address specific media phenomena with the help of all relevant multimodality-ready disciplines. Recurring meetings such as the International Conference on Multimodality (ICOM, last 2016 in South Africa) or the ACM International Conference on Multimodal Interaction (last 2017 in London) as well as events such as MODE 2015 (London, UK), Multimodality and Cultural Change (2015, Kristiansand, Norway), Transmediations! (2016, Växjö, Sweden), International Conference on Digital Media and Textuality (2016, Bremen, Germany), and more and more panels at other perennial conferences in the realm of communication studies, linguistics, and digital humanities, among others, dedicated exclusively to multimodality, show the need to discuss the problem-based approach. Furthermore, edited collections (partly evolving from these conferences, see for the most recent Norris and Maier 2014; Archer and Breuer 2015; Starc, Jones and Maiorani 2015; Wildfeuer 2015; Klug and Stöckl 2016), more and more introductions to the field (recently: Jewitt, Bezemer and O'Halloran 2016; Bateman, Wildfeuer and Hiippala 2017) as well as book series and journals around this topic (see, for example, the Routledge Studies of Multimodality, Multimodal Communication (de Gruyter), or the Journal of Multimodal Communication Studies (http://jmcs.home.amu.edu.pl)) help bring down the often spurious delineations and walls within the field of multimodal research. They show a good path of "rethinking" multimodality for the twenty-first century's media requirements: fostering scholarly exchange, assembling an array of high-quality multimodal research, and eventually integrating all that in a novel unified body of work that offers actionable theory, methods, and communication channels for modern multimodal research to follow.

This book follows the same path. It is the result of the Second Bremen Conference on Multimodality (BreMM15), dedicated to exploring the frontiers of multimodal research, which the two editors hosted at the University of Bremen, Germany, in September 2015. As a series of such events starts to form, the necessity for their proper documentation and dissemination becomes apparent, too. With each new event in the series (next: BreMM17 in September 2017), we discover new definitions, applications, iterations, operationalizations, and explorations of "multimodality" as a theory, concept, method, and phenomenon. In setting the stage for this volume of new studies in multimodality, we offer our working definition of the term, outline where the "new" most often lies, pay our respects to the timeless principles and thinkers that continue to shape multimodal research today, and offer greater insight into the individual chapters and their authors' motivations and contributions to the field.

2 Defining multimodality for the twenty-first century

Multimodality has been defined variously as a phenomenon (Iedema 2003), a theory (Jewitt 2009b), a methodological application (Jewitt 2009b), and later yet as a field of application (Jewitt 2014; see in particular the differences in the two versions of the *Routledge Handbook to Multimodal Analysis* (Jewitt 2009a)). It is natural to strive to put any field on solid ground with concrete definitions and sharp delineations. However, we argue that this is counterproductive in the case of modern multimodal studies. At the current stage of development and in keeping with the goal of a grand unified approach to multimodality, we propose to define it as a modus operandi for conducting research on human communication, both mediated and face to face. As such, it is more encompassing than a method and more palpable and pliable than a theory. Most importantly, it is inclusive and uniting rather than exclusive and divisive. While it falls short of the all-encompassing claim that "everything is multimodal," it does not forbid the application of multimodal methodologies or theories to any instance of creating and communicating meaning per se. With this, we follow a problem-oriented approach to the study of multimodality as a way of "characterising communicative situations (considered very broadly) which rely upon combinations of different 'forms' of communication to be effective," as recently suggested to be the most promising and effective way (cf. Bateman, Wildfeuer, and Hiippala 2017: 5).

Opening up multimodality in such a radical way may well seem like a risk. After all, being everything can devolve into being nothing rapidly. This is, however, a false fear. The requirements and expectations of media research in the twenty-first century are inherently multimodal: colleagues from a number of fields are developing sophisticated analytical schemes for analyzing the structures, compositions, intended as well as perceived meanings, and rigidity or fluidity of media content or human interaction alike. By conducting robust and reproducible research on such and other multimodal aspects of human communication, scholars, intentionally or inadvertently, join the global society of multimodalists. The sooner we create an explicit open space for them to exchange their expertise and bring it into the lap of multimodal studies, the stronger and better defined the discipline of multimodality will become.

The various forums and publications which we have singled out above have started doing the good work of bringing modern media studies into the multimodal universe. Nevertheless, this liberalization meets palpable pushback from tradition as well as habit. To ease the process, it may still be helpful to provide a structure, into which individual contributions can fit and, thus, increase their impact. The present volume reaches a conceptual-empirical fork in the road, and each direction carves out a niche in contemporary multimodal research. One concerns, for example, the theoretical basis multimodal researchers habitually use and involves the expansion, redefinition, or hybridization of key concepts. After more than twenty years of groundbreaking research on the inclusion of all semiotic resources and their interplay in the analysis, the definition of the semiotic mode, or the choice of theoretical frameworks from other disciplines, this intellectual exercise is most useful in breaking down disciplinary boundaries, forging new common terminologies, and building bridges between schools of thought. The first half of this volume is dedicated to examples of such conceptual multimodal work.

The other direction for expanding multimodal studies and building common bases for them in the present is based on empiricism that acknowledges the multimodal nature of modern communication and develops the means for analyzing it, especially in large batches, that is, multimodal corpora (cf. Bateman 2014a). This partly mimics the corpus-based methods of linguistics-inspired multimodal research, a strong existing strand today and a popular "role model" for many of the robust analytical constructs used as of late. On the other hand, large data volume processing, visualization, and mining are not only coming into play in the evolving area of digital humanities but are, by now, distinguishing features of all disciplines involved with big data. Much of modern media research keeps the general spirit of assembling, annotating, and analyzing swathes of these data and delves into the specific ways in which complex multimodal documents such as films, comics, or webpages relay information (see, e.g., O'Halloran et al. 2012; O'Halloran, Marissa, and Tan 2014; Hiippala 2015a, b). How that information is structured in its different forms and how it is perceived by recipients naturally belong in the field of multimodality. The second portion of this volume showcases examples of such empirical applications of multimodal principles, often in co-deployment with concepts from neighboring disciplines.

The "fork in the road" view of modern multimodal studies, therefore, should not be taken as a T-junction. Both paths outlined in this volume ultimately lead us to a broadened understanding of instances of human communication today, and both have to keep theory as well as application in mind when presenting their contributions. Therefore, the two approaches rather run parallel to each other, with frequent overlaps and a view to turning into a one-way, two-lane road.

3 Signposts on the modern multimodal highway

If you can bear with the road metaphor just a while longer, each road requires proper signage which all traffic participants can understand and follow in order to prevent accidents. The "signs" in this metaphor are the scholars whose work consistently informs new research on multimodal topics and whose theories and concepts anchor and direct scientific inquiry in the field. They have played a pivotal role in defining and delineating linguistics, semiotics, visual studies, and communication studies. Despite how seminal their work is, it does not forbid critical inquiry, rereading, and repackaging; in fact, it often invites these activities and, thus, moves our scientific understanding forward. The present volume is but one more illustration of the timelessness of their contribution.

First among those "old masters" is Charles Sanders Peirce (1931–58), whose sign theory continually informs the conceptualization of mediated meaning making and the various means of analyzing and classifying it. Apart from his popular threefold sign partition (not to mention the multitude of others he worked on), his work on translating meanings and incorporating a pragmatic component into communication studies largely influences our way of analyzing communicative artifacts today. In the same breath, we ought to mention M. A. K. Halliday and his functional grammar (1985; see also Halliday and Matthiessen 2004), which remains a strong pillar of modern multimodal research. It elevates the social aspect of the meaning-making process to the highest determinant of successful communication and imbues communication artifacts with different strata, which provide a reliable and commonly applicable basis, upon which analysis can rest.

A related lasting trend in multimodal research revolves around genre and the affordances that accompany generic demarcation. Exemplified well in the work of Michael O'Toole (2011 [1994]), James R. Martin (1984, 2001), and John A. Bateman (2008, 2014b), genre acts as a meaning-making facilitator and an anchor of scholarly interpretation all at once. Even when it is not explicitly mentioned, genre is an essential element of multimodal studies and shapes both the conceptual and methodological approaches researchers take to any given communication artifact. Many of the chapters in this volume work with representatives of different media genres and address the concept directly, demonstrating its centrality to the field. The inclusion of genre in our line of research will only grow in importance and impact as new forms of expression take hold thanks to technological innovation and global communication flows. Genre also serves as a manifestation of the central role context plays in creating, disseminating, and decoding communicative meanings (see, e.g., Müller 2008). It underscores the importance of shared social convention, which leads us back to Halliday's basic postulates as well as to modern takes on multimodality from the perspective of social semiotics (Kress 2010).

Gunther Kress and Theo van Leeuwen, together (2006 [1996], 2001) and solo (see, inter alia, Kress 2010; van Leeuwen 1999, 2005, 2011), are another lasting presence in multimodal studies. Taking a cue from Halliday, their work on various semiotics-inflected "grammars" of visual expression, color, or light among others has done a tremendous service to multimodality by bringing previously excluded communicative means into the mainstream. While they are less concerned with going through large corpora, Kress and van Leeuwen's drive to innovate and expand the boundaries of what multimodal research can (and should) concern itself with has left a large mark on the field, as is evident by the numerous citations, applications, and rereadings of their theories in this volume and others. Boeriis and van Leeuwen's (this volume) contribution is the latest testament to the lasting intellectual curiosity and impact of their work.

Perusing this list, it becomes apparent that these repeatedly cited and revered authors have been working in the spirit of multimodality all along, and their intellectual output is indeed an excellent guide for our new explorations. They embody the academic rigor, breadth, and sustainability that characterize the field. The diverse selection of contributions that makes up this volume is yet another testament to these qualities of multimodal inquiry. They are also the traits that make multimodality a strong candidate to become the dominant communication research paradigm of our time, thanks to both the resourcefulness of the big names above and their followers alike.

4 Tracing old trails, striking new paths

Our edited volume, following the "fork in the road" outlined above, is organized in two main sections: one dedicated to theoretical and methodological elaborations and the other to empirical applications and analyses. The collection illustrates the rich variety of topics modern multimodal studies tackle, starting from more abstract, theoretical constructs and principles and gradually moving to hands-on, practical applications of theory for the study of real-life mediated and interactive phenomena.

The theoretical section opens with a contribution on "Vectors" by Morten Boeriis and Theo van Leeuwen. The authors revisit and expand upon the meaning and scope of the term "vector" as it was first established by Kress and van Leeuwen in 1996. The authors go back to Rudolf Arnheim's (1974, 1982) work on the concept and offer a compelling rereading and expansion of it, which culminates into a system network to be used in further practical work with unpacking visual vectors' semiotic potentials. The chapter is especially remarkable as it features Theo van Leeuwen revisiting a concept he has developed with Gunther Kress in their emblematic work *Reading Images*—a rare occasion indeed.

Søren Vigild Poulsen continues the trend of rethinking established theoretical concepts with his treatment of "Transduction as Translation." He proposes a new way of classifying the transfer of meaning across semiotic modes at the example of Nordic myths presented in verbal and visual form, with different nuances and elements made salient or obscured in the transferal process. To frame the transduction-translation debate, Poulsen starts out with Kress's (2003, 2010) work on translation, critically approaches some of his followers' extensions of it, and draws inspiration from Peirce's (1931–58) three most popular sign types to create a new system for the classification of meaning translated across modes.

Morten Boeriis tackles stratification and offers an update of how it should be modeled in light of current development in multimodal media studies. The author relies on Halliday (1985), Kress and van Leeuwen (2001), and Hjelmslev (1961 [1943]) in his initial treatment of stratification and explores the ways an alternative modeling of the process can affect the ways in which we gauge the various realization contexts of the semiotic text. With its equally in-depth discussion of instantiation, the chapter has far-reaching implications for the mapping of the dynamic process of multimodal meaning making at different strata. It also raises the important question of how sophisticated modeling affects our understanding of complex processes in both theory and practice.

Hans-Jürgen Bucher brings in the perspective of reception studies and uses it as empirical support for his call to replace the traditional sign-based approach to multimodal analysis with one founded on discourse instead. With the help of data generated in eye-tracking studies as well as re-narration, questionnaires, and other verbal methods, the author makes the case for a theory of multimodality based on two other prominent communication-centered theories: social cooperation (inspired mainly by Paul Grice 1989) and communicative action (championed by Goldman 1970). Thanks to this approach, Bucher addresses multimodal meaning making in its instantaneous totality rather than as a sum of separate semiotic processes and, thus, offers a solution to the problems of compositionality and reception, which he sees as central to any theory of multimodality.

Arne Krause's contribution on functional pragmatics brings a German linguistic theory into multimodal studies and continues the social-action orientation of the previous chapter. The area of functional pragmatics, as developed by Ehlich and Rehbein (1982), takes a view of language as historical-societal action, and the author proposes the means to expand that perspective to nonverbal resources in a grand unified way, which has not been attempted previously in the German context. Relying on examples of multimodal interaction in teaching and learning, the author proposes an expansion of the HIAT system for the transcription of spoken language (Ehlich 1993) to include a more complete account of multimodal communicative actions as well. The results indicate that functional pragmatics' multimodal expansion is a promising way of transcribing complex interactions in a comprehensive and exhaustive manner.

Christopher Taylor's chapter "Audio Description: A Practical Application of Multimodal Studies" opens the volume's empirical and applied portion. It presents the findings of a EU-funded project dedicated to studying the practice of inserting narrated descriptions of screen-media artifacts' visual components as an aid to visually impaired recipients. Relying on theories and concepts that form the core of contemporary multimodal studies, Taylor offers an in-depth perspective into the often tough choices that audio description professionals face at every step of their task. Invoking various methods such as eye tracking or phasal analysis (cf. Baldry and Thibault 2006), the author arrives at a set of recommendations for good audio description practice, providing a fresh perspective on how movies mean for sighted recipients along the way, too.

Victor Lim-Fei and Serene Tan address the lack of systematic education in visual and multimodal literacy in their contribution "Multimodal Translational Research: Teaching Visual Texts." Building upon multimodal discourse analytical constructs originally developed by Kress and van Leeuwen (2006 [1996]) and O'Toole (2011 [1994]), the authors develop a comprehensive guide to teaching visual texts for secondary school students. It empowers both students and teachers to tackle the challenge of unpacking a multimodal artifact by providing them with a standardized vocabulary set that has not been implemented in the educational context before. Apart from offering a systematic view at the task, Lim-Fei and Tan's guide empowers students in their approach to understanding multimodal texts—another rare and valuable contribution where theory is directly and practically applied to occurrences of multimodality in daily life.

Hartmut Wessler and colleagues tackle the problem of detecting verbal and visual bias in their chapter "'Wikiganda': How Neutral Is Wikipedia?" They apply various automatic and semiautomatic methods in order to assess the levels of bias in the description of Greek populist party Syriza and its classification on the political spectrum across different language versions of its Wikipedia page. The authors also analyze the visuals that are chosen to represent Syriza and its leader Alexis Tsipras across language versions and track the changes of various depiction styles through time, tying them to relevant political events. Wessler et al. present a compelling case for the integration of automatic and manual methods for the effective study of multimodal communication, and they demonstrate successfully the effectiveness of multimodal analysis in dissecting media bias.

Carmen Daniela Maier offers a deep qualitative look into the multimodal construction of corporate heritage identity by a prominent British charitable organization. Offering a seldom seen multimodal escapade into the world of organizational communication, the author extends traditional approaches to analyzing heritage identity (Balmer 2013) with van Leeuwen's (2008) legitimation strategies and their realizations across various semiotic modes. Maier also considers intersemiotic complementarity (Royce 2007) as a further way of solidifying messaging about an organization's heritage multimodally. The visual, verbal, audio, and hypermodal heritage implementation strategies paint a rich multimodal picture of the NGO in question and pave the way for further analyses of not-for-profits' mediated identity constructions.

Finally, Yannik Porsché introduces contextualization analysis as a systematic way of scrutinizing discourse in face-to-face interactions. The author offers a combination of ethnomethodological interaction analysis and post-structural discourse analysis with contextualization analysis in order to achieve a complete mapping of multimodal interactions and their contextual embedding. Using a university panel discussion on the Bologna Process, Porsché illustrates the power of contextualization analysis through a comprehensive transcription procedure adapted from Jefferson (2004) and Mondada (2008), spanning utterances and their idiosyncrasies as well as gestures and gaze. Illustrated with rich transcriptions, the chapter examines the constructions of participants, epistemic territories, and discursive levels. The results demonstrate contextualization analysis' power of unveiling interaction dynamics and the resulting power relations and attempts at discourse fixation.

5 Conclusion: Multimodal studies, united in their diversity

As the chapter summaries above demonstrate, we have assembled here a volume spanning diverse multimodal traditions—regionally, disciplinarily, philosophically, and empirically. Our explicit goal in doing so was to amass proof of multimodality's border-transcending quality and to revisit some of its cornerstones with the idea of rearranging some or all of them, so that they line a common new path. To achieve that, we chose to anchor the volume firmly into the twenty-first century's fast-paced communication landscape where mediated and face-to-face interactions alike are faster, more complex, and constantly reinvented. From this perspective, a volume of fresh, cutting-edge theoretical elaborations and empirical applications was sorely needed, and we believe we have answered that call well with the book you are holding in your hands or reading on your device's screen.

The broader implications of this book—and of the BreMM conference series, of course—revolve around the idea of bridge building. The rest of this volume is, therefore, best assimilated with a few guiding questions in mind, namely: How does this approach relate to my concept of multimodality? What other disciplines share the same basic assumptions and rely on the same influential scholars or schools of thought? To what other media or generic examples can I apply this empirical approach? With the mindset of intellectual curiosity and discipline-transcendent thought, multimodality research will always stay current and relevant, and we see this book as a contribution to maintaining that open-minded approach alive in the field, creating new connections, adapting and streamlining old views to match modern communication's properties, and, ultimately, solidifying multimodality as a diverse yet united approach that brings unique insight and understanding of the world around us.

References

Archer, A. and E. Breuer (2015), *Multimodality in Writing. The State of the Art in Theory, Methodology and Pedagogy*, Amsterdam: Brill.

Arnheim, R. (1974), *Art and Visual Perception*, Berkeley/Los Angeles: University of California Press.

Arnheim, R. (1982), *The Power of the Center*, Berkeley/Los Angeles: University of California Press.

Baldry, A. and P. J. Thibault (2006), *Multimodal Transcription and Text Analysis. A Multimedia Toolkit and Coursebook*, London: Equinox.

Balmer, J. M. T. (2013), "Corporate Heritage, Corporate Heritage Marketing, and Total Corporate Heritage Communications. What Are They? What of Them?" *Corporate Communications: An International Journal*, 18 (3): 290–326.

Bateman, J. (2008), *Multimodality and Genre: A Foundation for the Systematic Analysis of Multimodal Documents*, London: Palgrave-Macmillan.

Bateman, J. (2014a), "Using Multimodal Corpora for Empirical Research," in C. Jewitt (ed.), *The Routledge Handbook of Multimodal Analysis*, 238–52, London/New York: Routledge.

Bateman, J. (2014b), "Developing a GeM (Genre and Multimodality) Model," in S. Norris and C. D. Maier (eds.), *Interactions, Images and Texts*, 25–36, Berlin, New York: de Gruyter.

Bateman, J. A., J. Wildfeuer, and T. Hiippala (2017), *Multimodality. Foundations, Research, Analysis. A Problem-Oriented Introduction*, Berlin, New York: de Gruyter.

Ehlich, K. (1993), "HIAT: A Transcription System for Discourse Data," in J. Edwards and M. Lampert (eds.), *Talking Data. Transcription and Coding in Discourse Research*, 123–48, Hillsdale: Lawrence Erlbaum.

Ehlich, K. and J. Rehbein (1982), *Augenkommunikation. Methodenreflexion und Beispielanalyse*, Amsterdam: John Benjamins.

Goldman, A. I. (1970), *A Theory of Human Action*, Enlgewood Cliffs, NJ: Prentice Hall.

Grice, P. (1989), *Studies in the Way of Words*, Cambridge, MA: Harvard University Press.

Halliday, M. A. K. (1985), *Introduction to Functional Grammar*, London: Arnold.

Halliday, M. A. K. and C. Matthiessen (2004), *Introduction to Functional Grammar*, 4th edition, New York: Edward Arnold.

Hiippala, T. (2015a), *The Structure of Multimodal Documents: An Empirical Approach*, London, New York: Routledge.

Hiippala, T. (2015b), "Combining Computer Vision and Multimodal Analysis: A Case Study of Layout Symmetry in Bilingual In-Flight Magazines," in J. Wildfeuer (ed.), *Building Bridges for Multimodal Research: International Perspectives on Theories and Practices of Multimodal Analysis*, 288–307, New York: Peter Lang.

Hjelmslev, L. (1961[1943]), Prolegomena of a Theory of Language, Wisconsin: University of Wisconsin Press.

Iedema, R. (2003), "Multimodality, Resemiotization: Extending the Analysis of Discourse as Multi-semiotic Practice," *Visual communication*, 2 (1): 29–57.

Jefferson, G. (2004), "Glossary of Transcript Symbols with an Introduction," in G. H. Lerner (ed.), *Conversation Analysis: Studies from the First Generation*, 13–31, Amsterdam/Philadelphia: John Benjamins.

Jewitt, C. (ed.) (2009a), *The Routledge Handbook of Multimodal Analysis*, London: Routledge.

Jewitt, C. (2009b), "An Introduction to Multimodality," in C. Jewitt (ed.), *The Routledge Handbook of Multimodal Analysis*, 14–39, London: Routledge.

Jewitt, C. (ed.) (2014), *The Routledge Handbook of Multimodal Analysis*, 2nd edition, London: Routledge.

Jewitt, C., J. Bezemer, and K. L. O'Halloran (2016), *Introducing Multimodality*, London: Routledge.

Klug, N.-M. and H. Stöckl (2016), *Handbuch Sprache im multimodalen Kontext*, Berlin, New York: de Gruyter.

Kress, G. (2003), *Literacy in the New Media Age*, London: Psychology Press.

Kress, G. (2010), *Multimodality: A Social Semiotic Approach to Contemporary Communication*, London: Routledge.

Kress, G. and T. van Leeuwen (2001), *Multimodal Discourse—The Modes and Media of Contemporary Communication*, London: Arnold.

Kress, G. and T. van Leeuwen (2006[1996]), *Reading Images—The Grammar of Visual Design*, 2nd edition. London: Routledge.

Martin, J. R. (1984), "Language, Register and Genre," in F. Christie (ed.), *Children Writing: Reader*, 21–30, Geelong: Deakin University Press.

Martin, J. R. (2001), "A Context for Genre: Modelling Social Processes in Functional Linguistics," in J. Devilliers and R. Stainton (eds.), *Communication in Linguistics: Papers in Honour of Michael Gregory*, 287–328, Toronto: GREF.

McLuhan, M. (1960), "Effects of the Improvements of Communication Media," *The Journal of Economic History*, 20 (4): 566–75.

Mondada, L. (2008), "Documenter l'articulation des ressources multimodales dans le temps: la transcription d'enregistrements vidéos d'interaction," in M. Bilger (ed.), *Données orales. Les Enjeux de la Transcription. Cahiers de l'Université de Perpignan. No 37*, 127–55, Perpignan: Presses Universitaires de Perpignan.

Müller, M. G. (2008), "Visual Competence: A New Paradigm for Studying Visuals in the Social Sciences?" *Visual Studies*, 23 (1): 101–12.

Norris, S. and C. D. Maier (eds.) (2014), *Interactions, Images and Texts: A Reader in Multimodality*, Hamburg: Walter de Gruyter.

O'Halloran, K. L., A. Podlasov, A. Chua, and K. L. E. Marissa (2012), "Interactive Software for Multimodal Analysis," *Visual Communication*, 11 (3): 363–81.

O'Halloran, K. L., K. L. E. Marissa, and S. Tan (2014), "Multimodal Analytics: Software and Visualization Techniques for Analyzing and Interpreting Multimodal Data," in C. Jewitt (ed.), *The Routledge Handbook of Multimodal Analysis*, 386–96, London, New York: Routledge.

O'Toole, M. (2011[1994]), *The Language of Displayed Art*, London: Leicester University Press.

Peirce, C. S. (1931–58), *Collected Papers of Charles Sanders Peirce*, in C. Hartshorne and P. Weiss (eds.), Cambridge, MA: Harvard University Press.

Royce, D. T. (2007), "Intersemiotic Complementarity: A Framework for Multimodal Discourse Analysis," in D. T. Royce and W. L. Bowcher (eds.), *New Directions in the Analysis of Multimodal Discourse*, 63–111, London: Lawrence Erlbaum Associates.

Starc, S., C. Jones and A. Maiorani (eds.) (2015), *Meaning Making in Text: Multimodal and Multilingual Functional Perspectives*, Basingstoke: Macmillan Palgrave.

van Leeuwen, T. (1999), *Speech, Music, Sound*, London: Macmillan Palgrave.

van Leeuwen, T. (2005), *Introducing Social Semiotics*, London: Psychology Press.

van Leeuwen, T. (2008), *Discourse and Practice. New Tools for Critical Discourse Analysis*, Oxford: Oxford University Press.

van Leeuwen, T. (2011), *The Language of Colour: An Introduction*, London/New York: Routledge.

Wildfeuer, J. (2015), "Bridging the Gap between Here and There: Combining Multimodal Analysis from International Perspectives," in J. Wildfeuer (ed.), *Building Bridges for Multimodal Research: International Perspectives on Theories and Practices of Multimodal Analysis*, 13–34, Frankfurt a.M.: Peter Lang.

2

Vectors

Morten Boeriis and Theo van Leeuwen

1 Introduction

The concept of "vector" derives from the study of art and design, but it has become a crucial component in Kress and van Leeuwen's (2006) account of the grammar of visual representation in which the authors merge concepts from the theory of art and design (especially from Arnheim 1974, 1982) with concepts from Halliday's functional-semantic theory of language (Halliday 1978, 1994). Art and design theory provides an account of the visual signifiers and linguistics—an account of the signifieds, of the broader cultural meaning system that, Kress and van Leeuwen (2006) argue, underlies not only language but also other semiotic modes. What parts of this cultural meaning system can then be expressed in what modes is, like the cultural meaning system itself, culturally and historically variable.

This approach risks selecting only those aspects of Arnheim's work that fit Halliday's linguistic theory, and has therefore occasionally been criticized as too logocentric. In this chapter, we approach vectors by rereading Arnheim and rethinking the theory of visual transitivity in that light. This will not, we will show, lead to a repudiation of Kress and van Leeuwen's work but to its enrichment and refinement.

2 Vectors

The term "vector" comes from the Latin verb *vehere* ("to carry") and the agentive suffix -tor. Its literal meaning is, therefore, "carrier." It was coined in the nineteenth century as a mathematical term for a force with a certain direction and a certain magnitude. The Russian-American philosopher Ushenko (1953) was the first to apply it to art, and for him it denoted a "force," a dynamic element, in visual composition. Its use in graphic design is summarized by Zettl (2013 [1973]: 132), who defines vectors as "created by simple lines or stationary elements that are arranged so that the viewer sees them as lines." Zettl distinguishes between "index vectors" that are "created by objects that unquestionably point in a certain direction" and "motion vectors" that "are created by an object that actually moves or is perceived as moving on a screen" (Zettl 2013 [1973]: 132). The concept of direction, and the relation to movement, is central in Zettl's definition, but it is a formal definition that provides no indication of the meaning potential of vectors.

Arnheim's account of vectors in *Art and Visual Perception* (Arnheim 1974) is also based on the physical and physiological forces, or rather, on their perceptual correlates and effects—vectors "counterbalance gravitational pull" (Arnheim 1974: 33) and thereby create "directed tension": "Motion, expansion, contraction, the process of growth—they can all manifest themselves as dynamic shapes" (Arnheim 1974: 416). One of his examples, a Toulouse-Lautrec drawing of a rearing horse, shows how "the shape of the object generates direction along the axes of its structural skeleton" (Arnheim 1974: 28) and demonstrates the importance of obliqueness in creating "directional tension." Oblique lines and oblique objects such as rearing horses are gravitationally unstable, in danger of falling, unless this is prevented by a force in a specific direction: "Oblique orientation is probably the most elementary and effective means of obtaining directional tension [and] with the mastery of obliqueness the child as well as the primitive artist acquires the main device for distinguishing action from rest—for example, a walking figure from a standing one" (Arnheim 1974: 426).

Unlike Zettl's, Arnheim's definition is, therefore, not formal but semiotic. It includes the signified: "Subject matter creates direction. It can define a human figure as advancing or retreating" (Arnheim 1974: 18). In short, vectors signify action. In *The Power of the Center* (Arnheim 1982), the semiotic role of vectors is developed further. Here the author explicitly states that "vectors control meaning" (Arnheim 1974: 153), and it is principally the account of the relation between "volumes" and vectors in this work that inspired Kress and van Leeuwen's theory of visual transitivity. As Arnheim puts it (1974: 154), "We shall distinguish between volumes and vectors, between being and acting."

Volumes are inert "visual objects," whose role in composition lies in their configuration, in the way they are distributed across the visual space. Vectors, on the other hand, work through the interrelation of directional forces and form the dynamic element. Both are present in all visuals. In the later work of Mondrian, for instance, "one can see such a painting as an agglomeration of rectangles and squares held together like a wall of bricks—that is, one can see volumes. Or one can see a grid of lines moving in vertical and horizontal directions—that is, vectors" (Arnheim 1974: 155).

But it is through the integration of these two components and through "their relation to another and their functions within the work as a whole" that meaning is made (Arnheim 1974: 153). Nevertheless, visual representations "vary in their ratio between volumes and vectors. . . . Certain early styles, e.g. that of the Easter Island statues, rely on compact volumes, and, at a more differentiated level, classical and monumental styles show a similar preference [whereas] others rely more strongly on vectors" (Arnheim 1974: 155).

In short, visuals can represent subject matter either as static or dynamic. But in Arnheim's account, this is not an "either-or" distinction—it is a matter of degree.

Kress and van Leeuwen's definition of vectors closely follows Arnheim. They define vectors as oblique lines with a sense of directionality and as issuing from and directed at volumes (Kress and van Leeuwen 2006: 59). These lines may be abstract (e.g., arrows) or formed by part or all of a depicted object (e.g., an outstretched arm, or Toulouse-Lautrec's rearing horse). And, as in Arnheim, they denote actions. The kinds of action will then depend on the objects that form the vectors, for example, rearing in the case of the rearing horse.

There are, however, also differences between Kress and van Leeuwen and Arnheim. For Arnheim, despite his emphasis on the importance of obliqueness, horizontal and vertical lines can also be vectors, as we have seen in the Mondrian example. For Kress and van Leeuwen, this can be so only if the lines include a signifier of directionality, for example, the arrowhead of an arrow. Although they do not categorically exclude the possibility that horizontal and vertical lines can be vectors they say that "a vector is formed by a (*usually diagonal*) depicted element, or an arrow" (Kress and van Leeuwen 2006: 74, emphasis added), they do not give "unusual" examples, and the point is not explored further. Unlike Arnheim, Kress and van Leeuwen also include what is called "magnitude" in the mathematical definition of vectors—vectors may be "amplified," made "bolder" (Kress and van Leeuwen 2006: 72).

Arnheim's volume-and-vector structures are then interpreted in terms of Halliday's transitivity theory (Halliday 1994), as realizing "narrative processes" analogous to Halliday's material processes (which are realized by clauses). Volumes are said to realize the participants engaged in actions, and vectors—the actions themselves, the "processes." The terms "participant" and "process" are thereby interpreted as semantic-functional categories that form part of the cultural meaning system as a whole and can be applied to different semiotic modes, even though different modes will realize them differently. In language, participants will be realized by nominal groups; in visuals by volumes. In language, processes will be realized by verbal groups; in visuals by vectors. The participants from which the vectors issue forth, or which, in whole or part, form the vectors, are then interpreted as Actors, the participants who *do* the actions; and the participants at which the vectors are directed are interpreted as Goals, that is, the participants to which the actions *are done*. In this way "Actor" and "Goal" also become "pan-semiotic" concepts, applicable to different modes, even though they will be realized differently in different modes—in language through word order (in an active clause the Actor precedes the process and the Goal follows it) and in visuals through the directionality of the vector. Kress and van Leeuwen's key example is an early-nineteenth-century engraving (as used in an Australian primary school social studies textbook), which shows, on the left, two men in early-nineteenth-century army uniform, their guns aimed at a group of Aboriginals seated around a fire on the right (see Figure 2.1). The outstretched arms and guns form an oblique line issuing from the men, who are, therefore, the Actor, and pointing at the group of Aboriginals, who are, therefore, the Goal of the process. Thus, the structure of the image is analyzed as analogous to that of a clause-like "The British (Actor) attacked (process) the

Figure 2.1 *The British used guns* (early-nineteenth-century engraving, cf. Oakley 1985; see also Kress and van Leeuwen 2006: 45).

Aboriginals (Goal)"—note that the caption represents the event differently by not mentioning what "the British" used their guns for.

Three further aspects of Kress and van Leeuwen's theory of visual transitivity should be mentioned, first the types of narrative process they recognize and the way these types are realized by different volume-and-vector configurations. In language, material process clauses can be transitive or intransitive. In a similar way, Kress and van Leeuwen (2006) argue, visual processes can either be "transactional" or "non-transactional," either include a Goal or not—it is, after all, perfectly possible to have an image in which we see people pointing their guns at something that is not shown in the image. Three different kinds of volume-vector relations are, therefore, possible: the transactional process (Actor + Process + Goal), the non-transactional process (Actor + Process), and what Kress and van Leeuwen (2006: 64) called an "event" (Process + Goal). An example of an event could be a map on which an arrow signifies the trajectory of an army, the action of "advancing," and a town, the destination of the advance. In such a case the Actor (i.e., the army) may be realized verbally, but the visual only "says": something is advancing toward the town.

Again following Halliday, Kress and van Leeuwen also distinguish different *kinds* of narrative processes, each characterized by a specific type of vector. Apart from actions and events, there are also reactions, realized by eyeline vectors, a kind of vector also recognized by Arnheim (1974: 28). Verbal processes are realized by a vector formed by the speech balloon protuberations that connect speakers to what they say (as enclosed in speech balloons) or by lines connecting speaker and speech balloon. Mental processes are realized by a vector formed by the usually diagonal thought bubbles that connect Actors to what they are thinking, as enclosed in thought balloons or realized by some kind of icon, for instance the "eureka" light bulb of Disney comic strips. Again, these types of processes are in principle pan-semiotic, realizable in different modes, albeit in different ways. But not all types of process can be realized in every mode. Halliday, for instance, also recognizes "behavioral processes," but these are not included in Kress and van Leeuwen's treatment of vectors.

Finally, Kress and van Leeuwen distinguish two broad classes of process. Besides "narrative processes," which represent doings of one kind or another, there are also "conceptual processes," which represent more or less fixed and permanent states of affairs and which can be recognized by the *absence* of a vector. If there *are* lines, they will be horizontal or vertical, for example, the lines in a tree diagram or the lines in certain kinds of "analytical" process, which do not signify some kind of action but "mean something like 'is connected to', 'is conjoined to', 'is related to'" (Kress and van Leeuwen 2006: 59). As we have seen, Arnheim also envisaged this possibility. But to him it was not an "either/or" distinction: visual works, he said, "vary in their ratio between volumes and vectors" (Arnheim 1982: 155). Some rely more on configurations of "compact volumes," others on "constellations of vectors." In Halliday's transitivity theory, a similar distinction can be made between, on the one hand, material processes, behavioral processes, verbal processes, and mental processes, and, on the other hand, relational and identifying processes, processes that assign attributes and identities to participants.

All these processes can be realized in all visual media and genres. Here, too, Kress and van Leeuwen follow Arnheim, who draws his examples not only from drawings, paintings, etc., but also from sculptures and buildings. Narrative processes occur in diagrams as well as images, as Kress and van Leeuwen (2006: 48) have shown in their discussion of "communication models" and in two-dimensional representations as well as in sculptures and product design (Kress and van Leeuwen 2006: 241–42). Analytical structures can be pictorial as well as diagrammatic, as shown in Kress and van Leeuwen's comparison of de Stijl paintings and city maps (Kress and van Leeuwen 2006: 91–92), and technology has, by now, more or less obliterated the distinction between photographic and non-photographic visual representation (though not between the photographic "look" and the "drawing" look). In short, the processes Kress and van Leeuwen describe are presented as, in principle, "pan-visual," even though, in a given context or period, not all may be realizable in every medium or genre.

3 Some questions

Kress and van Leeuwen's synthesis of Arnheim and Halliday has in many ways proved its usefulness in teaching and research. It helps us understand the role of composition in constructing visual representations. Nevertheless, in the course of using the theory, questions have arisen. We will discuss three of these in particular. The first is, do vectors always realize processes? Is the hookup between vectors and narrative processes as neat as we have represented it here?

The answer is "no," not even for Kress and van Leeuwen. In Halliday, and also in Kress and van Leeuwen, the transitivity system realizes the ideational metafunction, the semiotic function of constructing representations of the world. But vectors can also realize the interpersonal function, the function of creating symbolic interactions with the viewer, because eyelines (and gestures) may be directed not only to other represented participants but also to the viewer (Kress and van Leeuwen 2006: 117): "When represented participants look at the viewer, vectors, formed by the participant's eyelines, connect the participants with the viewer. . . . In addition there may be a further vector formed by a gesture in the same direction." Vectors can also realize the textual function, the function of "forming texts, complexes of signs which cohere both internally with each other and externally with the context in and for which they were produced" (Kress and van Leeuwen 2006: 43). In an analysis of a women's magazine article titled "A new breed of gold-diggers," Kress and van Leeuwen show how the article, on the left, describes the work of gold miners, and how a centrally positioned photo of workers smelting gold points toward a much larger photo, on the right-hand page, in which two women with safety helmets and fireproof clothing smile at the viewer: "The [central] photograph is tilted to form a vector that leads the eye to the photograph on the right" (Kress and van Leeuwen 2006: 180). Apparently, vectors can play a role in realizing all three of Halliday's metafunctions. Similarly, volumes can realize not only participants but also whole participant-and-process structures, as in Figure 2.2, or even major elements of a layout, whole stretches of text, and whole photographs. Flowcharts and similar diagrams provide further examples of textual uses of vectors. In Figure 2.2, for instance, the volumes realize the stages of a "life cycle." It is not the case that the volume "completed cars driven out" "*does something to*" the volume "cars driven for years. Regular maintenance." The arrows have a conjunctive function. They indicate the passing of time. They mean "and then," not "does to."

A second, related, question arises: Are actions and transactional always realized by a vector? Figure 2.3 shows people running into the sea. But their legs do not clearly form the kind of dynamic oblique lines that can be seen, for instance, in many sports photographs.

A third question relates to the multiplicity of vectors in many images. In Joseph Koudelka's memorable photo of a student challenging Soviet soldiers during the Prague Spring uprising of 1968 (Figure 2.4), the most salient vector is formed by the student's arm. The picture could therefore be paraphrased as, "The student (Participant: Actor) challenges (Process: Vector) the Soviet soldiers (Participant: Goal)." But it contains many other vectors: the soldier's legs, the eyeline of the pensive soldier on the right, the train tracks, the electric wires, and the line

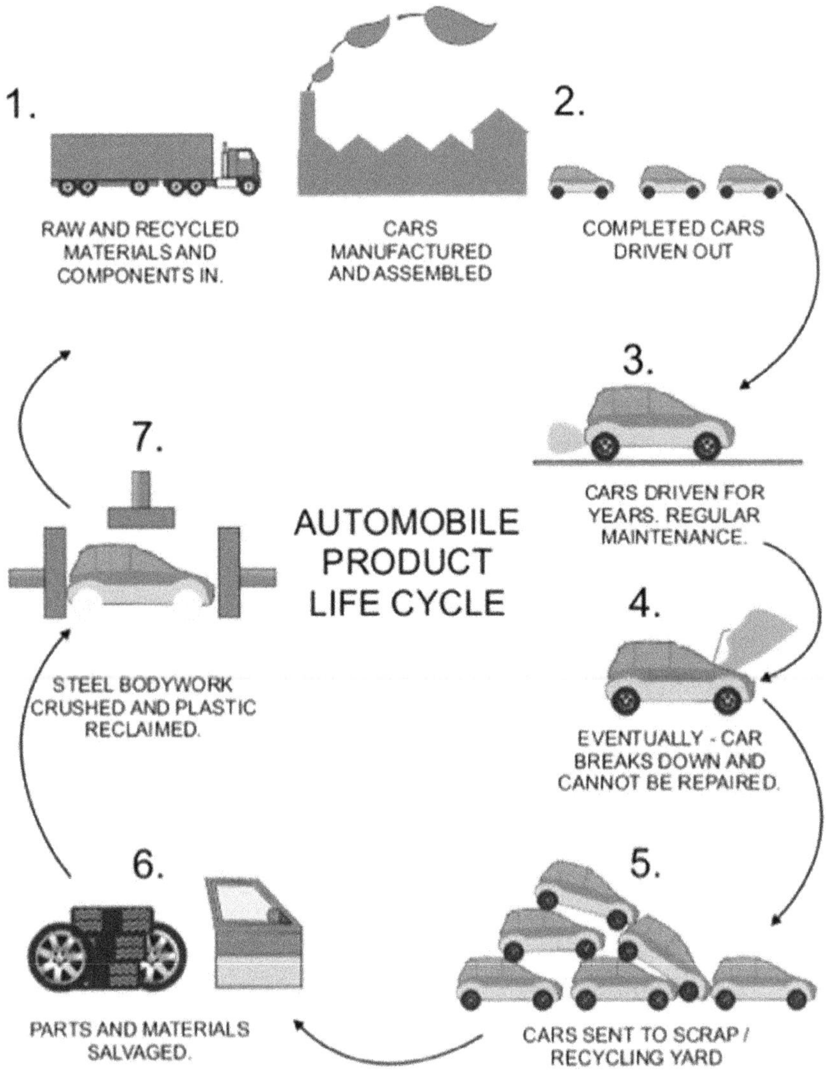

Figure 2.2 Automobile product life cycle (by permission from V. Ryan, World Association of Technology Teachers 2009).

formed by the treetops. We will return to this picture later. For the moment, we wish to ask: Have we too often confined the analysis of visual narrative structures to the most salient vector, and hence to simple structures that can readily be seen as analogous to clauses, whereas, if we take all vectors into account, a structure might emerge that is much less easily paraphrased as a clause?

With these questions in mind, we revisit Arnheim in the next section.

Figure 2.3 *Keep running II* (© Rachel Sian, Creative Commons License, https://www.flickr.com/photos/rachelsian/136141667).

Figure 2.4 Invasion by Warsaw Pact troops in front of the Radio headquarters. Prague. August 1968 (Joseph Koudelka 1968; © Magnum Photos).

4 Rereading Arnheim

A concept that is crucial in Arnheim's account of volumes and vectors, but not taken up by Kress and van Leeuwen, is "node," the place where vectors cross. Arnheim defines nodes as "constellations of vectors, which create centres of visual weight" (Arnheim 1982: 155) and stresses their semiotic importance, their role in "expressing the theme." "The constellations of vectors which I have called 'nodes' organize the visual matter and works of art into readable patterns. By watching what happens at these centres we understand what the work of art is about" (Arnheim 1982: 169). The vanishing point of central perspective is a node, because it is the place where the converging lines come together, but it does not always coincide with a volume. Both types of node can realize processes of a kind that has not been considered by Kress and van Leeuwen and cannot easily be related to linguistic processes.

4.1 Radiation and concentration

One of the types of node Arnheim discusses is formed by the "sheaves of concentric radii that emerge from a centre or converge toward it" (Arnheim 1982: 155). Here the vectors are not directed in one specific direction, as is the case with Toulouse-Lautrec's rearing horse, which rears up and toward the right, nor do they point at another volume, as in the case of Figures 2.1 or 2.4. In short, the process is not unidirectional, but multidirectional, and its semiotic potential lies in what it literally does—radiating outwards in all directions *from* a volume, or converging inwards from all directions *toward* a volume. In the former case, which we might term "radiation," the process is a nontransactional action—there is an Actor with multiple vectors issuing forth from it. In the latter case, which we might term "concentration," the process is a kind of event—there are multiple vectors which are all directed toward a Goal. Clearly these basic dynamics can be used for many more concrete processes. In one of Arnheim's examples, the severed head of John the Baptist radiates powerfully in all directions, hauntingly directed, not only at Salome but at the whole room (see Figure 2.5). It may not always be clear whether concentric radii should be interpreted as diverging or converging. The lines formed by the double tail of the mermaid in the Starbucks logo (Figure 2.6), for instance, could be interpreted as Starbucks expanding in all directions, across the globe, so to speak, or as everyone coming to Starbucks, from all directions—or perhaps both at the same time.

4.2 Finite and infinite settings

According to Arnheim (1974), the converging lines of central perspective also constitute a node, namely the vanishing point. This not only makes them dynamic but also the locus of a theme. But what kind of theme is it? Our first observation is that, although dynamic, these vectors do not signify an action. The theme they

Figure 2.5 *The Apparition* (Gustave Moreau 1876; © RMN-Grand Palais/René-Gabriel Ojéda).

express takes place in the background, as part of what Kress and van Leeuwen call a "setting" and describe as contrasting with the foreground action by being darker or lighter than the foreground, less saturated in color or more muted in tone, and represented in less detail (Kress and van Leeuwen 2006: 72). But what distinguishes settings in which we can see perspectival lines converging toward a vanishing point from settings in which this is not the case? Panofsky (1991), in his celebrated essay *Perspective as Symbolic Form*, and Arnheim stress that central perspective represents "infinity." In the Middle Ages, Arnheim explains, only God was infinite. In the Renaissance, the *world* became infinite, in science as well as in visual representation—and for many, this was threatening. Hence, "Artists tended to avoid spelling out the vanishing point" (Arnheim 1974: 297). Its location could be guessed but was blocked from view, as in Figure 2.7, where

Figure 2.6 Starbucks logo (© Starbucks USA).

a wall and a barred window obscure it, or as in Leonardo's *Last Supper* where the figure of Christ overlaps with the vanishing point, so that Christ, rather than a distant and unknowable future, dominates the picture.

Perspective is also related to time. Central perspective, Arnheim says, portrays space as "a flow oriented towards a specific end . . . a happening in time, a directed sequence of events" (Arnheim 1974: 298)—an end that is by definition unknown and unreachable. This can help us look again at Figure 2.4. The defiant gesture that symbolizes the uprising, making it look forceful and, therefore, potentially successful, obscures a vanishing point that is ominously wrapped in a haze of smoke, auguring the defeat that lies ahead.

4.3 Dynamic settings

In *Reading Images*, Kress and van Leeuwen indicate that vectors may be multiple, "to suggest the frequency or multiplicity with which the process occurs" (Kress and van Leeuwen 2006: 72). Arnheim also envisages this possibility when he discusses how "in the upright position a single vector almost always appears in the company of other parallel vectors" (Arnheim 1974: 214). This suggests the

Figure 2.7 *The Annunciation with Saint Emidius* (Carlo Grivelli 1486; © The National Gallery, London).

possibility of background vectors signifying not a specific action but a general sense of dynamism: a setting with neatly upright parallels, whether the trees in a forest or an abstract pattern, suggests a static, orderly world—a setting with oblique uprights suggests a more dynamic world; a world in flux. This can be a symmetrical, well-ordered world or an irregular world, which, depending on the context, could suggest disorder, randomness, wilderness, or playfulness (see Figure 2.8).

4.4 The human body

In several of our examples, the vectors are formed by parts or all of the human body, for instance by outstretched arms. As Arnheim remarks, "The human

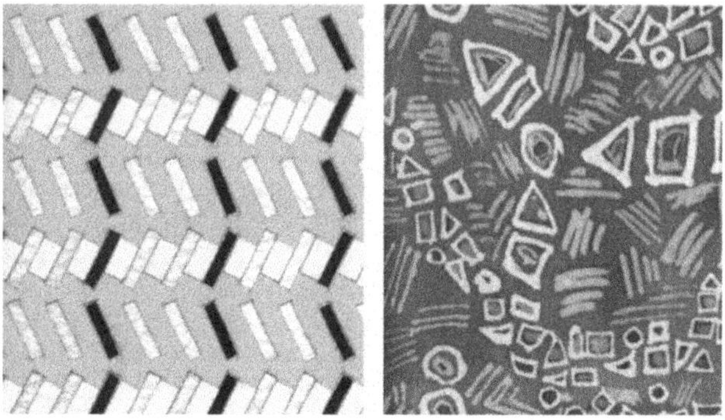

Figure 2.8 Regular and irregular dynamic patterns on wrapping paper designs.

figure is particularly well suited to show how volumes and vectors join efforts to compose expressive visual objects" (Arnheim 1982: 162). The torso then forms a node from which the limbs and the head branch out to realize actions. One of Arnheim's examples is *The Good Shepherd* (Figure 2.9), a fifth-century Italian mosaic in which "the sideways glance of the eyes, the oppositional gestures of the arms, which combine dominion and compassion, all characterize the Shepherd as the central node of the composition" (Arnheim 1982: 161).

In all this, Arnheim pays particular attention to the hands, "the most expressive nodes of dynamic action" (Arnheim 1982: 167), listing the kinds of action hands can realize: expressive actions such as wringing in despair, communicative actions such as pointing and beckoning, symbolic actions such as folding the hands together in prayer, functional actions such as grasping and holding objects, and signaling actions such as making a victory sign or counting on one's fingers. Although, for reasons of space, we cannot pursue this in detail, it clearly opens up the possibility of a visually rather than linguistically based inventory of actions and interactions, which, moreover, may be used ideationally, to represent actions, and interpersonally, to create mediated interactions—represented participants can, from within images, point at, beckon, or signal to the viewer, and so on. So-called "nonverbal interaction" has for the most part been studied in relation to face-to-face interaction, where gestures are actual movements, rather than what Zettl called "index vectors." But the patterns of actual body movements and the patterns of their representation in still images do not necessarily stand in a simple one-to-one relation and deserve further study.

Facial expression plays a particularly important role here. "Can a face be described as a node, as a constellation of vectors," asks Arnheim, and his

Figure 2.9 *Christ the Good Shepherd* (fifth century AD; Creative Commons License, http://diglib.library.vanderbilt.edu/act-imagelink.pl?RC=51106).

answer is yes. "It can if, as we must, we perceive the facial feature not as lifeless shapes but as vectors" (Arnheim 1982: 163). The eyebrows, the mouth, and the wrinkles and folds of the smiles and frowns that can contract or open up the face all clearly delineate vectors, resulting from the tension of facial muscles. The angle of a frowning vector, for instance, does not follow the line of the brow but goes downwards. There is, again, no space to explore this in detail here, but a few observations can be made. First, facial vectors do not have an Actor and a Goal in the sense in which we have so far described it—unlike eyelines, they neither emanate from a particular volume nor point at some other volume, even though in another sense, the person whose face is in question is the Actor, and facial expressions are communicative, expressing reactions in most cases. Second, they all center around the graded distinction between opening up (raised eyebrows, wide eyes, open mouth, etc.) and closing down (knitted eyebrows, half-closed eyes, pursed lips, etc.);—combinations of these, for instance pursed lips and widened eyes, can modify these expressions. In short, facial expression vectors realize a specific kind of reaction, different from the gaze. Their relation to verbal and mental processes also merits attention. As can be seen in Figure 2.11, the mouth can signify the action of speaking or singing without there being an utterance (in the form of a speech balloon). Facial expressions, similarly, can realize a mental process without there being a Phenomenon (in the form of a thought balloon)—although in combination with a transactional eyeline vector, it will be clear what the expression reacts to. A more detailed study of this kind of reaction is clearly needed and could perhaps start with the study of the systems practitioners have, implicitly or explicitly, designed for creating masks and puppets, cartoon characters, and so on.

5 The multiplicity of vectors: Two examples

Visual ideation not only draws on a wider range of processes and types of setting than have been foreseen so far, it is also a great deal more complex than linguistic ideation, which does not have an equivalent of the "node," the constellation of vectors. We will illustrate this with two examples. The first comes from Arnheim himself and is an analysis of the sculpture shown in Figure 2.10.

We will begin by quoting his brilliant analysis in a slightly abridged form:

> In seated figures, the main balancing centre is close to the floor, producing a solidly fastened focus, from which the various elements sprout upward in different directions. This is the basic constellation of vectors formed by the figure as a whole. . . . Within the vertical/horizontal framework the body rises as a powerful diagonal, kept by the bracing arms from falling backward, yet not reposing in itself. It hovers between rise and fall. Song, the subject of the work, is an action, and it is an upward-directed action as is revealed by the posture of the head, which is raised like that of a singing bird. This mighty sounding, sent from the ground to heaven, is reflected in the opening and spreading shapes of the figure as a whole, the wide spaces between the legs and between legs and torso. And we notice that under the impact of the rousing song the right foot lifts from the ground, as though the column of the vertical were pulled up and soaring. But this is only half the story. . . . Although wholeheartedly given over this song, our man is not addressing anybody. His eyes are tightly closed so that his attention is focused on the music from within. This counter-theme of concentration, indicated in the face, is spelled out more thoroughly in the other secondary node, the folded hands. They lock the brace of the arms, without whose grip the figure would fall apart. Severely contracted and safely tied to the ground, the sturdy figure this lets forth the powerful music without exploding or being torn from its mooring. (Arnheim 1982: 169–71)

The essential elements of a semiotic analysis are all present here—the signifying actions (bracing the arms, lifting the foot, raising the arms), whether they are

Figure 2.10 The Singing Man (Ernst Barlach 1928; © Scala Archives, DIGITAL IMAGE © The Museum of Modern Art/Scala, Florence).

transactional or not ("our man is not addressing anybody"), what kind of actions they are (the material action of bracing the arms versus the mental action of concentration, signified by the closure of the eyes) and the meanings which they realize. But most importantly, Arnheim analyzes the work as a network structure, a network of nodes, as a uniquely visual set of simultaneous volume-vector relations rather than as a clause-like structure. This does not disvalue the distinctions made in Kress and van Leeuwen's work, but it does bring out the complexity and specificity of visual syntagmatics.

Looking back at Figure 2.4 in a similar way, we can begin with the young student and his defiant gesture, his raised head and shouting mouth. But there is more. To his right, lower down, there are the soldiers' boots, pointing toward him. There is "directed tension" here, repressed force, as also in the tightly clasped hands of one of the soldiers. There is also the soldier, high up on the tank, looking at the student with a sullen, hostile expression. What strikes us first is the student's salient gesture of defiance and hope, but on closer inspection we can see that a storm is brewing—and the smoke in the distance, for the

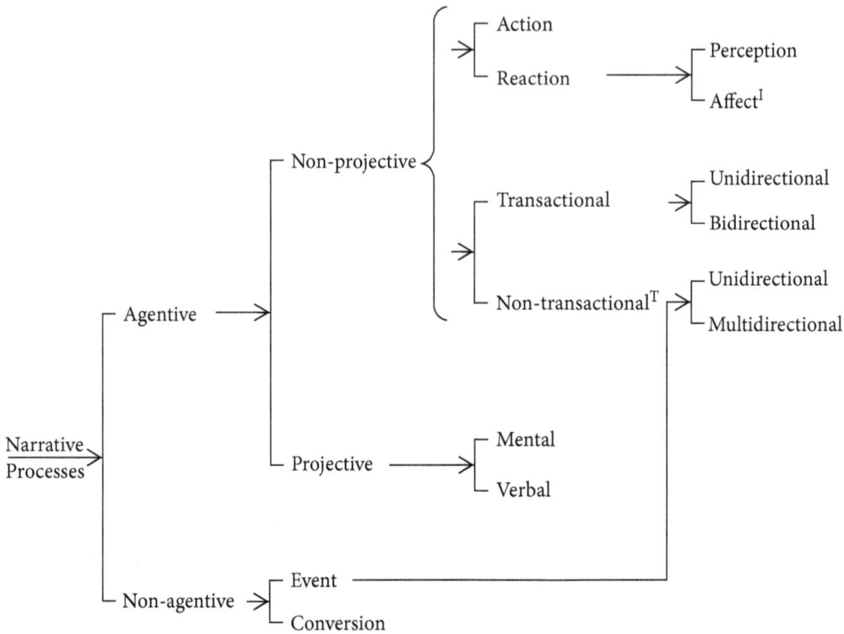

Figure 2.11 System network for narrative processes.

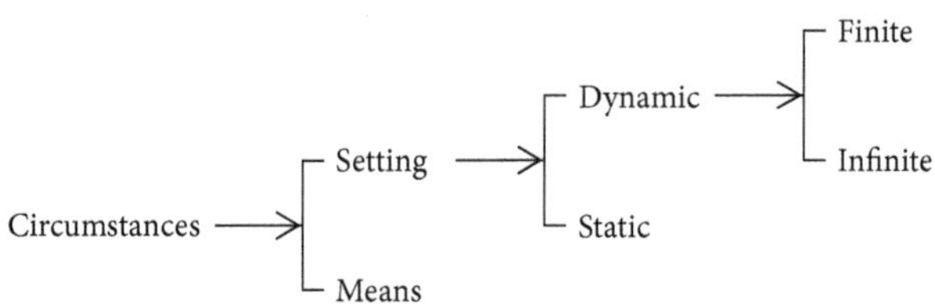

Figure 2.12 System network for narrative circumstances.

moment obscured by the central gesture, suggests what lies ahead, the crushing of the Prague Spring. All this is taken in at a glance, simultaneously present in a network of vectors (see Figures 2.11 and 2.12).

6 Vectors and processes

We now return to Figure 2.3, which represents the act of running without a clear, oblique "running vector," in contradistinction to Arnheim's notion that oblique lines distinguish action from rest, for example a walking figure from a standing one (Arnheim 1974: 426). Running is still signified here, however, but indirectly, through the "dynamic effects issuing from the environment" (Arnheim 1974: 239): we can see the vectors formed by the three bathers' arms as they try to keep their balance as well as the vigorous splashing of the water, and this leads us to conclude that they are running, rather than walking or standing. In short, there will always be a signifier but not always in a direct straightforward way. Running can be represented without having retinal presence as such, through what we might call a process of induced representation.

This does, however, make it important to distinguish between different kinds of lines. There are, first of all, the lines that form the contours of volumes. These do not necessarily have a vectorial function, even though they might, for example in the case of the rearing horse. Then there are vectors, and these will usually be oblique lines, but they do not need to be such in all cases. What they *do* need to convey, however, is what Arnheim called "gravitational pull" and "directed tension." An outstretched arm need not be oblique. It can be aligned with the horizontal or vertical axis. But we know that holding up your arm in this way requires force; that the arm is not at rest but performing an action. It *is* a vector. Finally, there are connectors, lines that connect volumes to form some kind of whole, in what Kress and van Leeuwen (2006: 87–104) call "analytical processes." But the distinction is not absolute. It is possible to give connectors a dynamic character, for instance by making them curved, or representing them as arrows, and, as we have already seen, it is possible to give vectors a relatively static appearance by aligning them with the horizontal or vertical axis. It is on the one hand important to realize that they still represent what is, in reality, either static or dynamic, but it is equally important to realize that these processes, regardless of whether they are static or dynamic in reality, are represented *as* static or dynamic—and that this is a significant part of their meaning.

7 Two system networks

We will summarize our discussion in the form of two system networks (see Figures 2.11 and 2.12), expanding the network of narrative structures and circumstances presented in Kress and van Leeuwen (2006: 74). It should be noted that they apply equally to direct and indirect (induced) representations, where the explicit vectors realize an implicit vector (as with the case of "running into the sea" in Figure 2.3).

The categories in these networks can be defined as follows:

Transactional action	An action involving an Actor, a process and a Goal, as realized by a vector issuing from the Actor (or being formed by part or whole of the Actor) and directed at the Goal
Nontransactional action	An action involving only an Actor and a process, as realized by one or more vectors issuing from the Actor (or being formed by part or whole of the Actor)
Event	An action involving only a Goal and a process, as realized by one or more vectors directed at the Goal
Conversion	A cyclical action in which the participants are the Goal of one action and the Actor of another, as realized by vectors connecting these participants
Unidirectional	A process with a single vector or more than one parallel vector
Bidirectional	A process with two vectors in which each of the two participants is the Actor of one action and the Goal of another (e.g., a reaction in which two people look at each other)
Multidirectional	A process with more than two vectors
Reaction	A process realizing a perception or an affect
Perception	A reaction realized by an eyeline
Affect	A reaction realized by facial vectors formed by wrinkles, eyebrows, eyes, and mouth
Verbal process	A process connecting an Actor to his or her utterance, as realized by a speech balloon vector
Mental process	A process connecting an Actor to his or her thought, as realized by a thought balloon vector
Setting	A circumstance of location distinguished from the process(es) it locates by decreasing color saturation, tonal contrast, and representation of detail
Dynamic setting	A setting containing multiple vectors
Static setting	A setting that does not contain vectors
Infinite setting	A setting that shows the lines of perspectival representation converging toward a vanishing point
Finite setting	A setting that does not contain the converging lines of perspectival representation
Means	An action formed, in whole or in part, by the tool with which an action is performed

It should be remembered that these distinctions describe a semiotic resource that can be used creatively in the production and interpretation of visual representations, not a code that must be adhered to. Meaning making is often an act of fusing what, in classification systems, are distinct categories. The networks should be seen as a systematic inventory of the options that the current state of the art of visual representation provides. If there are rules regulating their use, they are not inherent in the systems but created in and by specific social contexts (or technologies) to regulate specific practices of visual design or interpretation. While we hope that the new categories we propose here will be of use as tools for visual analysis, it should also be remembered that the use of these tools in the interpretation of visuals is ultimately a creative act, albeit one that must rest on arguments.

References

Arnheim, R. (1974), *Art and Visual Perception*, Berkeley/Los Angeles: University of California Press.

Arnheim, R. A. (1982), *The Power of the Center*, Berkeley/Los Angeles: University of California Press.

Halliday, M. A. K. (1978), *Language as Social Semiotic*, London: Arnold.

Halliday, M. A. K. (1994), *An Introduction to Functional Grammar*, 2nd edition, London: Arnold.

Kress, G. and T. van Leeuwen (2006), *Reading Images—The Grammar of Visual Design*, 2nd edition, London: Routledge.

Oakley, M. (1985), *Our Society and Others*, Sydney: McGrawHill.

Panofsky, E. (1991), *Perspective as Symbolic Form*, New York: Zone Books.

Ushenko, P. A. (1953), *Dynamics of Art*, Bloomington: Indiana University Press.

Zettl, H. (2013[1973]), *Sight, Sound, Motion Applied Media Aesthetics*, Belmont, CA: Cengage Learning, Inc.

3

The "Same" Meaning across Modes? Some Reflections on Transduction as Translation

Søren Vigild Poulsen

1 Introduction

The question I wish to explore in this chapter is whether transduction, or remaking meaning across modes of communication and representation, represents a category of translation. And if the answer is yes, why is this so? For example, if students remake a piece of music into a series of drawings, or if the written myths about the Nordic god Thor are transformed into paintings, images, or movies, does this constitute translation? Transduction is a cornerstone of Gunther Kress's theory of social semiotic multimodality and is widely used in multimodal literacy studies (e.g., Mavers 2011; Newfield 2009; Stein 2008). Kress (2010: 124) notes that translation, which "is the process of moving meaning," is a central semiotic process, and he operates with the subcategory of transduction (the remaking of meaning across modes) and transformation (meaning remade within the same mode). In his extensive work on transduction (e.g., Kress 1997, 2000a, 2005), Kress has shown how the translation of meaning across modes not only refers to how meaning in one mode can be remade in a different mode but also relates to the processes of redesigning meaning between various multimodal ensembles, genres, materialities, artifacts, and even cultures. Thus, transduction is a complex process involving all types of transmodal shifts in professional, artistic, educational, and everyday situational contexts. However, the definition of transduction as a translation remains rather vague. Little prior work exists on the nature of meaning translated across modes. Viewed from a translation perspective, transduction lacks precision in two ways: The first regards the types of meaning that might be translated across modes. This chapter addresses the tendency in Kress's (and others') work to focus on moving referential meaning, which I argue is not the whole story. Second, in Kress's definition of translation as a meaning-based concept, transduction also lacks precision about what makes this type of translation unique. In other words, when can transduction be labeled a translation and not another kind of semiotic process?

In this chapter, I present categories for analyzing the relationship between the specific meanings translated between modes—a dimension of the transduction concept, which seems to be underdeveloped in the existing literature. I build a twofold argument of what constitutes transduction by drawing on concepts from social semiotics as well as Peirce's semiotics. Following a review of selected social semiotic multimodal literature on transduction, the argument unfolds in a progressive manner divided into three sections. The first deals with how the three possible types of semantic, metafunctional, transductive meaning are constructed by drawing on concepts of stratification and metafunctions from systemic functional linguistics (SFL) and social semiotics. The second section offers a discussion of the relations of translated meaning inspired by Peirce's concept of icon, index, and symbol. The third section addresses critical issues in the argument.

2 Background

In terms of multimodality theory, several authors have focused on exchange and transference of meaning between modes (for an overview, see Newfield 2014). As Newfield (2014) notes, this phenomenon goes by different names in the literature, such as "transformation" (Kress 2000a), "transduction" (Kress 1997, 2000b, 2010; Kress and van Leeuwen 2001, 2006), "chains of semiosis" (Stein 2003, 2008), "transmodal moment" (Newfield 2009; Newfield and Maungedzo 2006), and "transmodal redesign" (Mavers 2011). Also to be included is Iedema's (2003: 41) "resemiotization," which describes how "meaning-making shifts from context to context, from practice to practice, or from one stage of a practice to the next." Despite differences of terminology, these designations all refer to approximately the same process, that is, how meaning is redesigned through modal transfers or shifts and how such redesigned meaning is manifested in different semiotic products (texts).

2.1 Kress's social semiotic perspective on transduction

Kress (1997) introduced the term "transduction" to describe how students remake meaning across different multi-semiotic artifacts, both inside and outside the classroom. He used the term to focus on how meaning is formed through rearticulation in different modes. However, Kress does not relate transduction to translation in a more general sense because his focus is on the learning context. He mentions briefly that translated texts have three metafunctions, but he does not elaborate further.

Kress (2010: 124) devotes a chapter to "processes for semiosis," and includes transduction as one central process. Kress presents transduction as "an absolutely common, constant, ordinary and profound process in everyday interactions" (Kress 2010: 125). Meaning remaking is guided by the interests of the sign makers and is regulated by the social context; it is created by the use of materials that aid the expression and conveyance of meaning (Kress 2010: 108). Furthermore, transduction changes the rhetorical or epistemological frameworks as a consequence of a modal shift because different organizational principles and cultural logics apply to the communicative use of the modes through which the transduction takes place.

For something to be classified as translation, a criterion of similarity between the original text and the translated text could be proposed as a prerequisite. But Kress (2010) does not talk of similarity in a transduction; however, he does reflect on "similarity" in general. As he notes, the "similarity" between meanings articulated in different modes may also be of a more abstract nature. For example, a book structure with an introduction and chapters and the architecture of an office building with an entry hall and different floors have the same meaning, which is an abstract order of elements (chapters and floors). The point is that meaning articulated in different modes can be abstract and vague, but nevertheless related. This being said, Kress's notion of transduction as a translation and possible criteria (such as similarity) for transduction remains ill-defined.

A second point to ponder is Kress's description of meaning remade in new semiotic artifacts by way of transduction. Although Kress links translation, and its subcategories transduction and transformation, to more general semiotic principles and provides a deep explanation of the transductive meaning-making process, he does not explicitly discuss translated meaning with regard to metafunctions. In fairness, it is possible to identify the meanings of Kress's analyses of different texts and artifacts. However, when reading the studies, it is clear that Kress primarily focuses on what would be the ideational meaning component. He does not distinguish between the types of translated meaning. Rather, he talks about meaning in more general and unspecified terms. Kress (1997: 134) points out that the texts produced in transduction have three metafunctions but not that the translated meaning could potentially involve more than one (ideational) meaning outcome as a result of the move from source text to target text, which is one of the claims of this chapter. I suggest that the meaning, which is translated across the modes, could very well be a closer link to the fundamental concept of metafunctions. While Kress has not explicitly done so, other social semioticians have introduced similar ideas.

2.2 Other social semioticians' perspectives on transduction[1]

Stein (2008) applies a multimodal framework for understanding specific activities in classrooms. In a series of case studies, she documents how this framework is useful. For instance, in a case study of the storytelling practices of Lungile, a 13-year-old Zulu-speaking female student, Stein, draws on Halliday's (1985) systemic functional grammar when analyzing meanings in written texts and uses Kress and van Leeuwen's (1996) visual grammar to analyze pictures. Stein shows convincingly that a social semiotic methodology is useful, but a reading of her work also indicates that the analysis tends to focus on metafunctional components in singular texts produced through acts of transduction. Thus, less attention is paid to theorizing about the relationship of meanings between texts in transduction. Furthermore, Stein applies the term "resemiotization" rather than "transduction." While the terms are, to some extent, related, I tend to view resemiotization as a more general semiotic practice in remaking meaning, whereas transduction is a subcategory of resemiotization.

Mavers (2009) discusses the advantages of a social semiotic multimodal framework in showing how students' use of dry wipe whiteboards is not only an activity but also a principled engagement of students with the taught subject and, thus, contains traces of serious semiotic work. She does mention metafunctions and demonstrates analytically that transduction concerns not only ideational but also textual meaning, focusing on textual patterns of coherence. She does not analyze the interpersonal metafunction, but one could easily do so. Furthermore, Mavers (2011) investigates the various ways of children's everyday text-making practices in drawing and writing. By employing a social semiotic framework, she demonstrates that the drawings and written texts children produce on a daily basis, often considered mundane, ordinary, imperfect, and unserious, are, upon closer inspection, intriguing and remarkable as semiotic work. They are texts produced in classroom routines that involve copying, remaking, and redesigning. To make her argument, Mavers employs social semiotic concepts. She does not, however, describe meaning remaking in the specific terms of stratification and metafunctions. In more general terms, she addresses the relations and orchestration of texts, that is, the interweaving of resources for constructing textuality from one text to another text in a new mode or medium. Thus, she presents an interesting view with respect to translation, albeit in general terms.

Newfield (2009), like Mavers, uses a social semiotic framework to analyze transduction as a key activity in classrooms. She mentions metafunctions and uses them to show and explain how changes of mode transform meaning. The focus of her work is on the transmodal shift; she coins the term "transmodal moment" for instances when meaning is translated from one mode to another. Here, less attention is given to the translation criteria. Focus is rather on the modal shift and the effects on meaning. Newfield makes a fine point about the profound impact of moving meaning across modes, but at the same time she questions the extent to which some meanings are "the same." To me, one hereby risks overemphasizing the difference between texts. For transduction to be classified as translation, the meaning has to be the "same," although redesigned or reshaped; otherwise, it is not a translation but rather a resemiotization. The problem also illustrates that there is a need to distinguish between content-substance (in SFL terms, meaning) and content-form (wording) in the discussion and understanding of transduction. Thus, while I agree that transduction reshapes and, thereby, transforms meaning, it does not seem quite enough to merit the use of the term "translation." Therefore, extending the work of Stein, Mavers, and Newfield, and with an ambition to use the social semiotic concepts of stratification and metafunction to talk more systematically about transductive meaning, my goal in the next section is to define translation across modes in a more systematic way.

3 An analytical framework

This chapter aims to present a twofold argument that seeks to make the descriptive analysis of translated meaning more precise.

3.1 Using social semiotic multimodality to define transduction

To build the first part of the argument, I need to describe a set of basic assumptions in a social semiotic multimodal framework.

3.1.1 Translation and text

The discussion of transduction starts from a simple notion of translation. According to Hatim and Munday (2004: 6), translation is "the process of transferring a written text from source language (SL) to target language (TL)." To avoid linguistic bias and to define translation as a social semiotic multimodal framework, I have tweaked the dictionary definition to the following working definition: translation is the semiotic operation of remaking the meaning of a source text into a target text. "Text" here is understood as any material articulation of semiotic work in a social setting (Kress and van Leeuwen 2001: 40). Thus, the translation is based on the meaning of a text and on the relation between meanings of the two texts. The working definition makes texts, not modes, the basic units of analysis. This (simplified) definition also makes it possible to refer to the source text as the starting point for a translation (within the same mode or across modes) into the target text, which is considered the outcome of this semiotic operation by a semiotic agent in a social context and related to a particular practice (e.g., a learning environment in an elementary school).

3.1.2 Mode and transduction

To elaborate the argument, it is useful to distinguish, as Kress (2010) does, between translation in the same mode (e.g., two written texts) and translation across different modes (e.g., written text and visual text). This distinction calls for a definition of a mode. Here we need to consider the problematic status of "mode" in the multimodal literature. I support the idea that further research will need to address this term (for an overview of the discussion, see Jewitt 2009). For the sake of the argument presented, a mode is pragmatically defined here as a means of communication and representation (e.g., written language, spoken language, image, gesture, gaze, posture, etc.). It is assumed that something can function as a mode if it is used by a social group to create and exchange meaning in a social setting. While most texts are designed as multimodal ensembles, in keeping with Kress's view, it is assumed that one mode in an ensemble functions as the primary means of communication, while other modes perform secondary, complementary functions. For example, in a written text, writing is the primary mode while color, typography, and layout are secondary modes; in a video, image is the primary mode, while speech, sound, and lighting are secondary. On this basis, and as an elaboration of the definition of translation, transduction is defined as the semiotic operation of remaking meaning of a source text in one mode into a target text in another mode.

3.1.3 Grammar

It is to be assumed that, in order to create meaning, a mode must have grammar, that is, a social group's shared understanding of a mode that displays regularities for the use of the mode, so that meaning in social settings may be made and enacted. Following Halliday (1978), the grammar of language (writing) is not a set of fixed rules but a resource for creating meaning. This notion of grammar has inspired social semioticians to think about other modes besides writing in a similar manner (e.g., Kress and van Leeuwen 1996, 2001, 2006; O'Toole 1994). It follows that images, gestures, dance, and other modes have grammar, however differently organized, that enable them to fulfill communicative functions similar to writing. Relating the notion of modal grammar to transduction means that the grammars of the source text mode and the target text mode provide various different resources that can realize meaning.

3.1.4 Transduction from a stratification perspective

Since meaning is the core issue in a transduction, it is useful to define this concept more precisely. This can be done by viewing meaning from the perspective of stratification. Following Halliday's (1978) description of language as a stratified semiotic system, one can distinguish between three levels (or strata) of language: semantics, lexico-grammar, and phonology or graphology.[2] The semantics and lexico-grammar concern the content side of language, while phonology or graphology falls on the expression side. In language, semantics is the meaning in layman's terms; lexico-grammar is wording (words and sentences); and phonology or graphology is distinct sounds and letters. The strata are related, which means that meaning is realized as words or sentences, which in turn are realized by sounds or letters. Given the assumptions about modes in the previous section, it follows that other nonlinguistic modes might be described in stratified ways similar to language. Take the mode of image as an example: the semantic meaning of an image is construed by the visual structures, which in turn are realized in terms of different means of visual expression. However, it is important to stress that nonlinguistic modes are not the same as language (given language priority); they are compatible because they are semiotic systems, and the language system is loosely used as inspiration for modeling the stratification of other modes (e.g., see still images: Kress and van Leeuwen 1996, 2006; moving images: O'Halloran 1994, Boeriis 2009; gestures: Martinec 2004). When parts of a mode as a semiotic system are instantiated, that is, manifested in a text, the notion of the three-leveled semiotic system can be used to clarify the concept of meaning, which is central to the present discussion of transduction and concerns the semantic strata of text. Thus, the first step toward an adequate description of transduction would be to define it as the remaking of the semantics of a source text in one mode into the semantics of a target text in another mode. So, from

Table 3.1 Stratification of a written source text and a visual target text

	Source text in writing	Target text in image
Content	Semantics (meaning)	Semantics (meaning)
	Grammar (wording *qua* words and sentences)	Grammar (visual structures)
Expression	Graphology (letters)	Visual expressions

this point on, the meaning of a transduction will be referred to as semantics. Table 3.1 illustrates the transduction of semantics between a written source text and a visual target text, both displayed as stratified semiotic systems. The gray fields indicate the level related to transduction.

3.2 Three kinds of transduction

This chapter argues that transduction might not only be about the translation of a subject matter from a source text in one mode to a target text in another mode, although this would be a commonsense way of considering this semiotic operation. We translate meaning about a particular aspect of or idea about the world from one text to another. However, there can be other types of meaning that are translated, either in conjunction with the translation of some meaning about the world or on their own, thereby creating a translation that is not based on subject matter but on other types of meaning. To elaborate this point, we may apply the concept of metafunctions (Halliday 1978, 1985; Kress and van Leeuwen 1996, 2006; Kress 2010). Metafunctions are three interrelated dimensions of meaning of relevance to the use of a mode (e.g., language or image), within a specific context. In SFL, social semiotics, and social semiotic multimodality theory, there are three overall metafunctions: ideational, interpersonal, and textual. Ideational metafunction is the representation of aspects of the experienced world, inside and around us; interpersonal metafunction is the establishment and maintenance of social relations between social actors; and textual metafunction is the capacity to form texts (i.e., the discursive organization of text units of internal and external coherence to fit the situational and cultural context). All texts have these three metafunctions, so in a transduction, both the source text and the target text can be individually described according to them. For the purpose of this chapter, the concept of metafunctions is used slightly differently to refer to the type of meaning in a transduction that is translated from the source text to the target text. The suggestion is that the translation of texts, in different modes, can be based on the remaking of one or several metafunctions of a source text and not necessarily on the ideational type. All metafunctions must, therefore, be considered when discussing translatability. In Halliday's words:

> "Translation equivalence" is defined in ideational terms; if a text does not match its source text ideationally, it does not qualify as a translation. . . . For precisely this reason, one of the commonest criticisms made of translated texts is that while they are equivalent ideationally, they are not equivalent in respect of the other metafunctions—interpersonally or textually, or both. (Halliday 2001: 16)

Taking this approach, combined with the stratification perspective introduced in the section above, it is possible to distinguish between three types of transductions, that is, the remaking of three types of semantics from a source text in one mode into a target text in another mode. A transduction may include the translation of one or several metafunctions on the semantic stratum, that is, experience, interpersonal relations, and textual coherence.

3.2.1 *Ideational transduction*

This type of transduction translates the ideational semantics about world experience of a source text, that is, processes, actions, states, or events, into the target text semantics. Ideational transduction means that the source text and target text represent the same subject matter. In other words, they share representation for the "same" content. Since the source and target texts have different grammars, it follows that they have different ways of encoding ideational semantics, thus providing different resources for representing experience. Examples of ideational transduction can be seen in the myth of Thor's visit to the giant Utgard-Loki in the comic book *The Travel to Utgard-Loki* (Madsen 2014/1989), particularly in the scene where Thor attempts to lift Utgard-Loki's cat (Figure 3.1). The text reads:

> Then spoke Utgard-Loki: "It is obvious that your strength is not great. Will you try more contests?" Thor says: "I may as well have a try at still more contests. But it would seem strange to me, if I were at home with the Aesir, if such drinks were reckoned small there. But what game do you want to offer?" Then Utgard-Loki replies: "It is for young lads to lift up my cat off the ground. But I would not know how to mention such a thing to Thor of the Aesir if I had not seen that he is a lesser man that I have been told." Then a grey cat ran out onto the hall floor, rather a big one. Thor put his hand down under its belly and lifted it up. But the cat arched its back and as he stretched up his hand the cat raised one paw. Then

Figure 3.1 Drawing of Thor attempting to lift Utgard-Loki's cat.
Source: Madsen (2014/1989:85). Courtesy of the author.

> spoke Utgard-Loki: "It went just as I expected. The cat is rather large, but you are short and small." (Sturluson 2012: 71)

In the comic book, both processes and participants represented in the written text are remade, not in words, but in images.[3] By stretching his arm under the belly of the cat, the vector of Thor's arm realizes the process of lifting, and both

Thor and the cat are represented participants in the frame (Madsen 2014/1989: 85–88).[4] Another example of the transduction of the lifting process could include Niels Hansen Jacobsen's (1891–92) sculpture *Thor and the tomfoolery in Utgard.*

3.2.2 Interpersonal transduction

This transduction translates the interpersonal semantics, that is, the social relationship between sender and receiver (and the represented ideational content), in the source text to the interpersonal semantics in the target text. Thus, the transduction relates to the extent to which the "same" interpersonal semantics may be remade across texts in different modes. Again, an example from the myths about Thor: In Hymir's Poem in *The Poetic Edda*, Thor goes fishing with the giant Hymir, and Thor hauls up the World Serpent, a sea monster. In the scene, Thor is portrayed in a heroic manner:

> Then very bravely Thor, the courageous one,
> pulled the gleaming serpent up on board.
> (Larrington 1996: 81)

This glorified portrayal of Thor is remade in numerous paintings, for example, by Füseli (Figure 3.2) or by Frölich (Figure 3.3). Of course, this glorified portrayal of Thor is made with different means in writing and images; writing employs attributes to the participant (Thor is brave and courageous), whereas images use vertical perspective (in both images, Thor is seen from a low angle; thus, he looks powerful and superior) and the modality markers of light, colors, and size are used (Kress and van Leeuwen 2006). Nevertheless, these images are examples of remade interpersonal meaning because this strand of meaning is not only a question of the truthfulness or credibility of the ideational content but also of the intersubjective coloring of the representation, that is, how the sender perceives the content, and how the implied receiver is prompted to understand the same content (Boeriis 2009: 269).

In other words, it is about the ways that the writer or painter evaluates things or ideas and how the reader or viewer is invited to share this perception. As with ideational transduction, interpersonal transduction involves reflecting on the ability of different modes to enact the "same" semantic meaning about the social relationships between the sender and receiver of the texts. And again, the grammars of the different modes involve encoding the interpersonal semantics differently, from which it follows that the "same" interpersonal semantics are also transformed when remade across modes.

Figure 3.2 *Thor battering the Midgard Serpent* by Henry Füseli, 1790. Courtesy the Royal Academy of Arts, London.

3.2.3 Textual transduction

This transduction refers to the translation of the source text's textual semantics into the textual semantics of the target text, and thus, the textual organization of the source text is remade in the target text. An example of textual transduction can be found in the lines of Hymir's poem, which follow the quote in the example of interpersonal transduction. As Thor hooks the World Serpent, his grasp of the hammer is foregrounded, when he strikes the beast:

> With his hammer he struck the head
> violently, from above, of the wolf's hideous brother.
> (Larrington 1996: 81)

In writing, "with his hammer" is accentuated (in SFL terms, placed in the THEME), thereby making this part marked. The foregrounding of Thor's

Figure 3.3 *Thor, Hymir, and the Midgard Serpent* by Lorenz Frölich.
Source: Rydberg (1906: 915).

hammer is translated from the written text to, for example, an illustration in the Icelandic manuscript of *Saemundar and Snorra Edda* (1760) by Ólarfur Brynjúlfsson (Figure 3.4).

The foregrounding is remade differently in the visual mode: Thor's hand holding the hammer is placed at the top and the left of the painting—or, in Kress and van Leeuwen's (2006) information value terms, Given, and Ideal—and is made salient with the contrasting colors of brown and white. This textual metafunction runs opposite to the painting by Henry Füseli (Figure 3.2), where the hammer fades into the background. Elaborating on the previous table, the semantics of both source and target texts may be subdivided; the dotted lines indicate the three proposed kinds of transduction (see Table 3.2).

Figure 3.4 Illustration of Thor catching the World Serpent.

Source: Brynjúlfsson (1760/1999: 93). Courtesy of the Danish Royal Library.

Table 3.2 Ideational, interpersonal, and textual transduction from a written source text to a visual target text

	Source text in writing	**Target text in image**
Content	Semantics (meaning) – experience – interpersonal relation – textual coherence	Semantics (meaning) – experience – interpersonal relation – textual coherence
	Grammar (wording *qua* words and sentences)	Grammar (visual structures)
Expression	Graphology (letters)	Visual expressions

4 Transductive relations between source and target texts

Up until this point, it has been assumed that transduction, as an act of translation, is about the transfer of the semantics of the source text in one mode to the semantics of the target text in another mode, so as to make them "the same." This assumption of sameness is well known in translation studies, as the Danish translator Budtz-Joergensen points out:

> From a layman's point of view, one characteristic of a translation is that even though it is written in a different language, it is still somehow the same as the original. In Translation Studies, this relationship of sameness is termed equivalence, a "key issue throughout the history of Translation Studies" (Manfredi 2008: 65). However, it remains one of the most hotly debated issues in the discipline. Generally, scholars disagree about two things. First, whether or not it makes any sense to insist that source and target texts must be "the same." Second, what is the exact nature of this sameness. (Budtz-Joergensen 2015: 12)

In terms of language translation, the issue of sameness or equivalence also applies to transduction. Thus, I suggest that it is still possible to describe transduction with reference to a notion of sameness or meaning equivalence, which is a key concept in translation studies (Holmes 1972/1988; Catford 1978). However, we also need to address the opposite of sameness, that is, to what extent the meaning of source and target texts differs because transduction transforms meaning, thereby construing a complex relationship between the texts involved in transduction. Furthermore, translation, from a commonsense perspective, suggests some influence on or causal relation of the source text to the target text, and it seems important to capture these relations because they constitute transduction as translation. This reflection expounds upon issues mentioned in the introduction to this chapter. It is my view that we need to elaborate on the relationship between the meanings of the source text and target text to clarify what makes transduction a subcategory of translation and not merely a modal shift. When is something a matter of transduction, and, just as importantly, when is it not? So far, it has only been argued that transductive meaning (semantics) may be metafunctional. However, this still leaves room for the discussion of what allows us to see transduction as translation.

4.1 Types of semantic relations

The second part of my main argument in this chapter claims that transduction is constructed by a set of semantic relations between a source text and a target text. Transduction, in a prioritized order, (i) rests on a similarity of semantics in the source text and the target text; (ii) relies on a difference of semantics in the texts; and (iii) includes a causation of the source text's semantics on the target text's semantics. On this basis, the nature of transduction can be described, and it becomes possible to determine different degrees of translation across texts in various modes. The tripartition of similarity, difference, and causation is loosely based on Peirce's sign theory of icon, symbol, and index, respectively, but they are renamed to avoid unintended confusion of terminology because the object of study is not the representational function of a single sign, but rather the semantic relationship between source and target texts. It is important to point out that since I rename Peirce's sign types and use them to signify different semantic relationships, I have omitted all reference to the vast research on Peircean translation studies (see Petrilli 2003; Gorlée 2015 for an overview). Furthermore, given that this is the only part of Peirce's semiotics upon which I draw, other parts of Peirce's semiotics are left out.

Peirce defines the triad of icon, index, and symbol as follows: "A sign refers to an object by virtue of an inherent similarity ('likeness') between them (icon), by virtue of an existential contextual connection or spatiotemporal (physical) continuity between sign and object (index), or by virtue of a general law of cultural convention that permits sign and object to be interpreted as connected (symbol)" (Peirce 1931–1958: CP 2.247–9).

In discussing the semantic relation between the source and target texts, I have been inspired by Peirce's three sign types of icon, symbol, and index.[5] To be precise, Peirce's triad was used in a slightly changed order. The first type, the icon, inspired the category of similarity because an icon represents its object based on similarity, and one semantic relationship between the source and target texts may also build on the similarity of the represented content. The symbol inspires the category of difference because a symbol represents its object based on convention or habit, and the second type of semantic relationship between the source and target text may build on convention. In other words, the semantics of the two texts are not similar; they display different semantics that are only related by virtue of a translation convention. In thinking about the types of semantics relations (between a source and a target text), I was superficially inspired by Peirce's negative definition of a symbol as not being based on sameness (nor on causation) but on social convention and, thus, being an arbitrary relation (Peirce 1931–58: CP 5.73; EP 2.460-1; MS 7.72, MS 939.45-7).

If we were to take this negative definition and transfer it to the relations between the semantics of the source and target texts, and if the first category of semantic relation rests on a notion of sameness or equivalence, then the second category could be defined negatively as a relation to be inequivalent or different. By contrasting similarity with difference, one can also create a useful dichotomy between these two terms, which allows for thinking of translation as a continuum with similarity at one end, difference at the other, and a particular transduction of semantics somewhere in between these extremes. Finally, the index inspired the category of a causational semantic relation because an index represents its object based on a relation of (physical or causal) connection (Peirce 1931–58: 1.372, 2.281-5, 5.75). For Peirce (1931–58, EP: 2.291), "An Index is a sign which refers to the Object that it denotes by virtue of being really affected by that Object." For me, that concept of affection is useful to include in the discussion of translating semantics between the texts in different modes because transduction emphasizes the question of the nature and extent of transferred meaning. Thus, the third type of semantic relationship between source text and target text is built by the effect (or impact) of the source text's semantics on the target text's semantics. The target text is said to be affected by the source text. In sum, what I will use from Peirce is not his sign types as such but the ways in which a sign (representamen) might represent its object by either similarity, convention, or causation.[6]

4.1.1 Relation of semantic similarity

All translations, including transduction, presuppose a similarity (or equivalence) of semantics between the source text and target text; if there is no similarity, we would not dub it "translation." Thus, a semantic relation based on the similarity of the source and target texts is a prerequisite for any transduction. In the first place, this relationship would constitute transduction as translation. If there is no similarity between the semantics of the two texts, no transduction will have taken place. In the previous section on transduction and metafunction, the given examples of ideational, interpersonal, and textual transduction of semantics (e.g., actions, participants, social relations, textual structures) would be classified as similarity-based relations because the semantics of the source text is the "same" as the semantics of the target text. To the extent that it would be considered to make sense that the "same" semantics can be remade across modes with different affordances, a relation of similar semantics between the source and target texts would be the determining factor for categorizing transduction as a kind of translation.

4.1.2 Relation of semantic difference

It is evident that, in transduction, something will remain untranslated and specific to the instantiated meaning made in a particular text at a given time and place. Thus, there will also be elements that are not the same in the source and target texts. The second type of semantic relation could, therefore, be based on the difference of semantics in the source and target texts. While the transduction builds on some things being the "same" in some regard (similar or analogous), with regard to the text, there are also arbitrary elements that are not similar or analogous, what we might call a motivated relation. In this sense, the semantic relation is only established by convention. For instance, written myths about Thor mention his red beard, but not all images do. In Winge's painting

Figure 3.5 *Thor's fight with the Giants* by Maarten Eskil Winge 1872.
Courtesy: Nationalmuseum, Stockholm.

(Figure 3.5), the thunder god is depicted without a beard (other examples could include Wilhelm Petersen's drawing *Thor and Hymir fish for the Midgard Serpent* (1938), or Jack Kirby's Marvel comic figure [1962]). Thus, the attribute of Thor without a beard is not motivated by the written (source) texts; in the Norse mythology, Thor is described as having a red beard.

On a more general level, this second type of semantic relation could include all three metafunctions that are not translated from the source text to the target text. With these two categories, similarity and difference, it becomes possible to construct a scale of translation, going from absolute similarity (e.g., both the source and target texts construe the exact "same" semantics) to absolute dissimilarity (e.g., the semantics of the source text is completely different from the semantics of the target text), thereby, raising the question of whether the transduction qualifies as a translation at all. An avant-garde transduction with no immediately recognizable content from the source text in the target text would be an example. Between these two extremes, a transduction can be evaluated, and on this basis, we might interpret the degree or extent of the transduction.

4.1.3 Relation of semantic causation

It is further argued that beyond similarity and difference, transduction includes causation between the source text and the target text. It must be made clear that causation builds upon and presupposes a similarity of semantics in the two texts. Take as an example Thor's visit to Utgard-Loki:

> They walked on to Utgard and saw a castle standing on some open ground and had to bend their heads back to touch their spines before they managed to see over it. There was a gate across the castle entrance. Thor could not manage to get up over it and they squeezed between the bars. They saw a great hall, went in and saw there men that were big enough. . . . Then a long trencher was fetched and put on the hall floor, full of meat. (Sturluson 2012: 69)

The written text mentions the castle, the gate, the great hall, and the trough (the Danish translation also mentions rows of giant men sitting on benches). All these things are also represented in the drawings of the comic book *The Travel to Utgard-Loki* (Madsen 2014/1989); thus, a semantic relation of similarity between the source and target text is established. However, additional elements such as skulls and a fireplace are added to the frames (Madsen 2014/1989: 74–75). The trough that the men eat from is decorated with ornamentation not mentioned in the reference of the written text. The relation of semantic causation is established when the written text prompts certain elements to be translated from the drawings, and the drawer is a sign maker who elaborates on the depiction of the castle hall.

In this particular example, the written text, as a source, is more general than the drawing, but the opposite could also be the case; the written text might mention more activities, people, and so on than the image depicts. Thus, the relation of causation introduces the question of whether or not the semantics from the source text is elaborated or restricted when translated to the target text. So, in a discussion of translation across modes, it is not just the modes that afford certain qualities to the representation of the various elements; one has to include an element of causation into a description of the semantics that are transferred. The source text has influenced the remake of the meaning found in the target text, too. By translating across modes, which constructs a relation of semantic similarity, the sign maker also creates a semantic relation of causation, that is, a motivated cause-effect between (some semantic elements of) the source text and the target text. Defined this way, the source text is the cause, and the target text is the effect by way of transduction. All types of translation, within the same mode or across modes, include this dimension of causation; it must, therefore, be added to the analytical framework.

5 Discussion

Two things are to be discussed in this section: (1) possible combinations of metafunctional transductions and the semantic relations; (2) transductions, modes, and grammar.

5.1 Combining metafunctional transductions and semantic relations

Once the concepts of three kinds of transduction and the three types of semantic relations in a transduction have been presented, one might ask if the concepts can be combined to form more categories and subcategories of transduction to elaborate on the analytical framework.[7] It seems unproblematic to suggest that the first semantic relation, based on the similarity between the source text and the target text, can be metafunctional, that is, the transduction might be established by virtue of equivalent ideational, interpersonal, and/or textual semantic meaning in the source text and the target text. The examples in Section 3.2, illustrating that the transduction of meaning (semantics) can be metafunctional, also illustrate three semantic relations that are based on similarity. In other words, the ideational semantics of the source text and that of the target text are the same, and this may also be the case with, respectively, the interpersonal semantics and textual semantics of the two texts (these subcategories of transduction, of course, presuppose that the modes of source and target texts both have the ability to create all three metafunctions). Thus, one needs to broaden the perspective on the dimensions of meaning that might be translated across texts in different modes. If we go back to the critique of Kress's view on transduction in the introduction and literature review of this chapter, it follows that transduction can include the remaking of more than the ideation, or, in popular terms, the content, of a source text in a target text. We can now refine the argument by saying that translation is constituted by a semantic relation of similarity and that similarity can possibly divide into all three metafunctions.

The second relation, based on different semantics in the source text and the target text, may also be centered on one or more metafunctions. We might look at this as transduction constructing a different semantics from the source text to the target text in terms of the ideational, interpersonal, or textual metafunctions. It also follows that the similarity and difference relations go hand in hand; if the transduction is constructed by only one of the three metafunctions, such as in similar ideational semantics, then the other metafunctions must logically be interpersonally and textually different (to the extent that it is possible to distinguish analytically between the three). Furthermore, it is not possible to discuss metafunctional subcategories and relations of similar and different semantics in the source and target text without also addressing the key question of what constitutes a transduction as a translation, which I have addressed in this introduction. As Budt-Joergensen (2015) points out, in layman's terms, a translation would suggest that the meaning of a target text is similar to that of the source text (cf. Section 4). Halliday (2001) makes a similar point when he writes that translation is usually talked about in terms of ideational equivalence, although it need not exclude the other metafunctions. To continue this line of thinking, I would suggest that an argument about transduction as translation can be made. For example, provided that a relation of similar or equivalent semantics in the source and target text is established, the transduction may count as a translation. However, second, the similar semantics equivalent needs to be of an ideational nature for the transduction to classify as translation. So, while the concept of metafunctions is useful to show that translated semantics might include more than ideation, the concept of similar semantic relations is also needed to pinpoint the nature of transduction as a semiotic process. A transduction must translate the ideational semantics to count as a translation, and this implies a relation based on semantic similarity between the source and target texts. Defined this way, we can also answer the question of whether the semantic relation of similarity of transduction needs to involve ideational, or whether it can just be interpersonal or textual. The answer may be that the ideational dimension of transduction constitutes the core. If this core is missing, if the transduction only focuses on similar interpersonal or textual meaning but different ideational meaning, then we might not classify it as translation across modes. This would be the logical consequence of splitting semantic equivalence into metafunctions. Therefore, on this basis, one is able to present criteria for transduction.

The previous discussion leaves the question of the relation between semantic causation and its three transduction types, that is, the impact of the source text on the target text in relation to metafunctional semantics. Due to the way this relation has been described in the previous section, and in view of the argument that transduction requires a similarity between the source and target texts' ideational semantics, it seems reasonable to assume that this relation depends on and is especially related to ideational semantics. In a discussion of a semantic cause-and-effect relation between the source text and the target text, the causation would revolve around what is ideationally represented in the source text, and its extent. In other words, the causal relation between the source and target texts' semantics presupposes a relation of similarity between the texts' ideational semantics. Take again the example of the description of the hall in Utgard, which Thor goes to visit. In the written myth, the hall is sparsely described as a room in which giants sit around the fire, eating, and drinking. In the comic book, the hall is also represented, so a relation of similar ideational semantics between source and target texts is established. However, the comic book visualizes the hall in greater detail (thereby technically also establishing a relation of different ideational semantics). So the comic book adds elements to the ideational semantics. However, I would claim that these elements are motivated by the semantics of the source text, and the example, therefore, illustrates a kind of causational relation of the (written) source text onto the (visual) target text. The added elements still make a coherent whole and are related semantically to a mythical universe of giants and goods.[8]

Additionally, if we leave out semantic causation, we miss out on an important dimension of transduction that relates to the sign maker's available apt resources for meaning making and the social setting that influences the transduction. In the section about the semantic relation of causation, I mentioned that causation also raises the question of expanding or restricting the represented content, when the semantics of the source text is remade in the target text. The same question is even more relevant when connecting the relation of semantic causation in a transduction to the three metafunctional transductions. I would claim that this question cannot be addressed properly without also bringing in the concepts of modes and their grammars that provide different potentials for meaning making. Discussing if the (metafunctional) semantics of a source text have an influence or effect on the remaking of the semantics in a target text depends to a large degree on the meaning potentials that the grammar of different modes provides. The different modal grammar determines whether semantics is expanded or restricted when the semantics of the source text in one mode are remade in the target text in another mode and, thus, influences the relation of semantic causation that the source text might have on the target text. This leads to the second issue to be discussed.

5.2 Transduction, modes, and grammar

I can only briefly discuss this issue, although it could potentially further qualify the framework by discussing similarities and differences of grammatical resources within different modes involved in a transduction. Take the remaking of a representation of activities (such as Thor attempting to lift Utgard-Loki's cat) from writing to image as an example. This remaking concerns ideational transduction, which means that the source text and the target text represent the same subject matter. Since the source and target texts have different grammars, it follows that they have different ways of encoding ideational semantics, thus providing different resources for representing experience.[9] By looking more closely at grammatical similarities and differences, we must also address issues of transduction in a more systematic way. Using the transduction from a written text to a visual text as the principal example, both have the ability to construe experience by modeling reality as made up by processes (Halliday 1994: 106). In writing, experience is primarily represented and organized as different kinds of processes by the grammatical system of transitivity. In this system, experience is construed by three elements: the process itself, that is, an activity or goings-on; the participants, that is, the subjects or objects involved; and the circumstances and the environment. Given the modal affordances of writing, processes are realized in the verbal group, participants in the nominal group, and circumstances in the adverbials and prepositional phrases. In images, the transitivity system is realized by the visual grammar (i.e., visual structures), which may be either narrative or conceptual (Kress and van Leeuwen 2006). In narrative structures, the processes are typically realized by vectors (i.e., lines representing some goings-on), and participants, that is, entities in the picture, circumstances, or participants who "could be left out without affecting the basic proposition realized by the narrative pattern" (Kress and van Leeuwen 2006: 72). In the conceptual structures, participants are related to part-whole structures, realized in the image by the alignment of elements, indicating some taxonomy or attribution. It seems possible to argue that the meaning in writing, by way of transitivity, relates to images that also include transitivity, although the grammar of the two modes is differently organized and manifested. However, they are both semiotic systems able to construe ideational meaning by virtue of resources from the transitivity subsystems in the written mode and the visual mode. The point is that there may be certain similarities and differences between the grammatical resources of different modes. For example, in certain cases, both writing and images have similar resources, while in others the same meaning is realized by means of different resources, as we have seen in the example of interpersonal transduction. In the cases where modes provide analogous grammatical resources, such as systems of transitivity, we must ask how compatible these resources are.

To this comes the question of semantic causation because even if two modes have the resources to instantiate the "same" semantics, they may be at different stages of development, phylogenetically speaking. Written language, for instance, is more developed in terms of speech (communicative) functions, while images are less developed, even if they are also able to express a kind of speech functions ("images acts"). According to Halliday (1994), language has four functions of utterance (statement, question, offer, and demand), while images, according to Kress and van Leeuwen (1996), have two (offer and demand).[10] It follows that you can translate meaning between language and images; it might be the "same" semantics, but a transformation of meaning is still taking place at the grammatical stratum. Returning to the previous issue of discussion, if we are to discuss semantics relations and metafunctional transductions, we must also address the issue of relating modes and their different grammar more directly to one another.

6 Conclusion

This chapter discusses the social semiotic concept of transduction as a category of translation. It presents arguments on three types of transduction – ideational, interpersonal, and textual – based on metafunctions. Furthermore, it suggests the translational character of transduction described in three relations of semantics based on, respectfully, similarity, difference, and causation. The chapter discusses how the concepts can be applied to form categories of transductions to advance the analytical understanding of transduction as a translation of meaning across texts in different modes. For further research, the chapter invites a test of the suggested types of transduction and semantic relations on a larger text corpus to validate the adequacy of the analytical concepts. Also, we need a number of detailed comparative descriptions on how the specific resources of different modes transform meaning moved between modes. This way, we can gain more insight into the nature of transduction.

Acknowledgments

The author would like to thank Professor Denise Newfield of the University of the Witwatersrand, South Africa, and Associate Professor Morten Boeriis of the University of Southern Denmark for their helpful suggestions and advice on this book chapter. The author would also like to thank the editors for their support during the writing process. The arguments and opinions in this chapter are the sole responsibility of the author.

Notes

1 One could also include a Systemic Functional Linguistics approach to translation studies, but this literature only addresses translation of written language. Manfredi (2014: 23–24) mentions SFL studies of intersemiosis (to use Jakobson's term of translation from language to nonverbal semiotic systems), but excludes these from his handbook on state-of-the-art Systematic Functional Translation Studies.

2 In the following, I leave out the lexico-grammar term and just refer to the term "grammar" or grammatical system. I also refer to this older version of Halliday's (1978) language theory, well knowing that it has been further developed, but keeping to this simpler version to make my point.

3 There can, of course, be more and other sources to the selected target texts; in the case of *The Travel to Utgard-Loki* and other comic books in the Valhalla book series, the stories are inspired by many different Nordic myths and legends. The point is not to trace the direct inspiration source, but to show that equivalent semantics between written and visual texts can be found and described.

4 I must emphasize that the chapter does not present a systematic, exhaustive analysis of the referenced texts; only snippets are mentioned to illustrate analytical points.
5 It is crucial to understand that Peirce's three sign types serve only as inspiration. The three types of semantic relations suggested are not iconic, symbolic, and indexical in the traditional Peircean sense, although they are the main source of the conceptualization of three different possible semantic relations construed in a transduction.
6 To illustrate the three semantic relationships, the focus is on ideational transduction when talking about semantic relations and how the ideational semantics of the source text relates to the ideational semantics of the target text.
7 One could also include the relation of stratification and metafunction, but it has already been discussed in Section 3.1.3. In the same section, I have also briefly discussed the possible translation of meaning at the levels of grammatical and expression strata.
8 The degree of causation, of course, also depends on cultural knowledge and conventions of style and genre in the comic book medium.
9 A comparative analysis and discussion of grammars exceed the scope of this chapter; in the present discussion, a selected comparison will be presented.
10 One, of course, has to be careful not to uncritically use linguistic terms to describe images (for a discussion of the language inheritance on visual social semiotics, see Boeriis 2009).

References

Boeriis, M. (2009), "Multimodal Socialsemiotik og Levende Billeder [Multimodal Social Semiotics and Living Images]," PhD diss., Odense: University of Southern Denmark.

Brynjúlfsson, O. (1760/1999), *Saemundar og Snorra Edda*. Digital version by Svindborg, B., Royal Danish Library.

Budtz-Joergensen, P. (2015), "Translation with Systemic Functional Linguistics," MA diss., Roskilde University, Roskilde.

Catford, J. C. (1978), *A Linguistic Theory of Translation: An Essay in Applied Linguistics*, Oxford: Oxford University Press.

Gorlée, D. L. (2015), *From Translation to Transduction: The Glassy Essence of Intersemiosis*, Tartu: University of Tartu Press.

Halliday, M. A. K. (1978), *Language as Social Semiotic: The Social Interpretation of Language and Meaning*, London: Edward Arnold.

Halliday, M. A. K. (1985), *An Introduction to Functional Grammar*, 1st edition, London: Edward Arnold.

Halliday, M. A. K. (1994), *An Introduction to Functional Grammar*, 2nd edition, London: Edward Arnold.

Halliday, M. A. K. (2001), "Towards a Theory of Good Translation," in E. Steiner and C. Yallop (eds.), *Exploring Translation and Multilingual Text Production: Beyond Content*, 13–18, Berlin/New York: Mouton de Gruyter.

Hatim, B. and J. Munday (2004: 6), *Translation: An Advanced Resource Book*, London: Routledge.

Holmes, J. S. (1972/1988), "The Name and Nature of Translation Studies," in J. S. Holmes (ed.), *Translated! Papers on Literary Translation and Translation Studies*, 67–80, Amsterdam: Rodopi.

Iedema, R. (2003), "Multimodality, Resemiotization: Extending the Analysis of Discourse as Multi- Semiotic Practice," *Visual Communication*, 2 (1): 29–57.

Jewitt, C. (ed.) (2009), *The Routledge Handbook of Multimodal Analysis*, 1st edition, London/New York: Routledge.

Kress, G. (1997), *Before Writing: Rethinking the Paths to Literacy*, London: Routledge.

Kress, G. (2000a), "Design and Transformation: New Theories of Meaning," in B. Cope and M. Kalantzis (eds.), *Multiliteracies: Literacy Learning and the Design of Social Futures*, 153–61, Melbourne: Macmillan.

Kress, G. (2000b), "Multimodality," in B. Cope and M. Kalantzis (eds.), *Multiliteracies: Literacy Learning and the Design of Social Futures*, 182–202, Melbourne: Macmillan.

Kress, G. (2005), "Gains and Losses: New Forms of Texts, Knowledge, and Learning," *Computers and Composition*, 22 (1): 5–22.

Kress, G. (2010), *Multimodality: A Social Semiotic Approach to Contemporary Communication*, London: Routledge.

Kress, G. and T. van Leeuwen (2001), *Multimodal Discourse: The Modes and Media of Contemporary Communication*, London: Arnold.

Kress, G. and T. van Leeuwen (1996), *Reading Images: The Grammar of Visual Design*, London: Routledge.

Kress, G. and T. van Leeuwen (2006), *Reading Images: The Grammar of Visual Design*, 2nd edition, London: Routledge.

Larrington, C. (1996), *The Poetic Edda*, Oxford: Oxford University Press.

Madsen, P. (2014/1989), "The Travel to Utgard-Loki," in Peter Madsen (ed.), *Valhalla—Den samlede saga 2* [Valhalla—the collected saga 2], Vol. 2, 58–135, Copenhagen: Carlsen.

Manfredi, M. (2008), *Translating Text and Context. Translation Studies and Systemic Functional Linguistics* (Volume I: Translation Theory), Bologna: Dupress.

Manfredi, M. (2014), *Translating Text and Context. Translation Studies and Systemic Functional Linguistics* (Volume II: From Theory to Practice), Bologna: Dupress.

Martinec, R. (2004), "Gestures that Co-Occur with Speech as a Systematic Resource: The Realization of Experiential Meanings in Indexes," *Social Semiotics*, 14 (2): 193–213.

Mavers, D. (2009), "Student Text-Making as Semiotic Work," *Journal of Early Childhood Literacy*, 9 (2): 141–55.

Mavers, D. (2011), *Children's Drawing and Writing: The Remarkable in the Unremarkable*, New York: Routledge.

McArthur, T. and R. McArthur (1993), *The Oxford Companion to the English Language*, Oxford: Oxford University Press.

Newfield, D. (2009), "Transmodal Semiosis in Classrooms: Case Studies from South Africa," PhD diss., University of London, London.

Newfield, D. (2014), "Transformation, Transduction and the Transmodal Moment," in C. Jewitt (ed.), *The Routledge Handbook of Multimodal Analysis*, 2nd edition, 100–13, London/New York: Routledge.

Newfield, D. and R. Maungedzo (2006), "Mobilising and Modalising Poetry in a Soweto Classroom," *English Studies in Africa*, 49 (1): 71–94.

O'Halloran, K. L. (2004), "Visual Semiosis in Film," in K. L. O'Halloran (ed.), *Multimodal Discourse Analysis: Systemic Functional Perspectives*, 109–30, London/New York: Continuum.

O'Toole, M. (1994), *The Language of Displayed Art*, London: Leicester University Press.

Peirce, C. S. (1963–66), *The Charles S. Peirce Papers*, The Houghton Library, Cambridge, Mass.: Harvard University Library Microreproduction service. (The Peirce manuscripts are available on 30 reels of microfilm, in-text references are to MS).

Peirce, C. S. (1998), *The Essential Peirce, Selected Philosophical Writings*, Volume 2 (1893–1913), Peirce Edition Project. Indiana University Press, Bloomington and Indianapolis, IN. (in-text references are to EP, followed by volume and paragraph number).

Peirce, C. S. (1931–1958), *Collected Papers*. Vol. 1–8. Vols. 1–6 edited by Charles Hartshorne and Paul Weiss; vols. 7–8 edited by A. W. Burks. Cambridge, MA: Harvard University Press. (in-text references are to CP, followed by volume and paragraph number).

Petrilli, S. (ed.) (2003), *Translation, Translation*, Amsterdam, New York: Rodopi.

Rydberg, V. (1906), *Teutonic Methólody—Gods and Goddesses of the Northland*, Vol. III, London: Norrcena Society.

Stein, P. (2003), "The Olifantsvlei Fresh Stories Project: Multimodality, Creativity and Fixing in the Semiotic Chain," in C. Jewitt and G. Kress (eds.), *Multimodal Literacy*, 123–38, New York: Peter Lang.

Stein, P. (2008), *Multimodal Pedagogies in Diverse Classrooms: Representation, Rights and Resources*, London: Routledge.

Sturluson, S. (2012), *The Uppsala Edda*, in H. Pálsson (ed.), trans. A. Faulkes, London: Short Run Press Limited.

4

Modeling Multimodal Stratification

Morten Boeriis

1 Introduction

The semiotic landscape has changed dramatically over the last decades, and one of the most qualified attempts to address these changes has taken place in the multimodal branch of social semiotics, which is the focus of this chapter. Multimodal theory in its social semiotic approach has taken a point of departure in the more general semiotic axioms of Halliday's linguistic theory, which has led to many fruitful insights into multimodal meaning making (Andersen et al. 2015). It is interesting to consider how social semiotic multimodality has branched out to numerous areas of communication and has, to a great extent, managed to encompass very diverse semiotic resources inspired by semiotic insights from a fundamentally linguistic theory. Nevertheless, it seems there has been less focus on discussing and revisiting the basic axioms of systemic functional linguistics from a multimodal perspective, on exploring what the multimodal understanding provides to and demands from the theoretical axioms—that is, discussing what may have to be altered to encompass a plurality of semiotic systems. Metafunction has been the point of departure for quite a few of these social semiotic endeavors into multimodality (for instance Kress and van Leeuwen 1996, 2001; O'Toole 1994; Baldry and Thibault 2005), and it has been seen as a defining feature of a semiotic system, or mode as it has been called (Kress and van Leeuwen 2001: 21–22; Kress 2010: 79), in that it handles three simultaneous functions in meaning making, namely ideational meaning, interpersonal meaning, and textual meaning (Kress 2010: 87). Some of the other axioms in Halliday's theory have been subject to less attention in the multimodal literature and may need to be readdressed as a consequence of the multimodal turn of the theory. One of the axioms that has perhaps been the most underexposed in multimodality is that of stratification, which stems from "the idea that language consists of different coding levels" (Taverniers 2011: 1102). In a multimodal perspective this means the distinction between different levels of encoding in semiotic systems along the stratification axis of the two sides of the traditional sign dyad of "expression" and "content."

This article discusses the idea of stratification as a multimodal phenomenon by proposing a remodeling of stratification, based on a rereading of Hjelmslev's, Halliday's, and Kress and van Leeuwen's takes on the concept in the light of multimodality. The point of departure for this article is a (rather radical) multimodal approach where the traditional understanding of modes as separate systems as well as analytical abstractions has been abandoned because the phenomena we encounter are never monomodal; therefore, in contemporary communication, analytical abstraction is a less productive heuristic, since the semiotic resources are increasingly more detached from their artisanal origin.

2 Hjelmslev's stratification of language

First, we turn to Hjelmslev, who describes language as an abstract system and takes a perspective from inside language itself, much along the lines of Saussure. To Hjelmslev, the sign is pure form, which can be seen as quite different from a more traditional understanding of a sign as something that stands for something else (Hjelmslev and Uldall 1967: 27). Hjelmslev isolates the sign to the abstract form and states that "the sign is a two-sided entity, with a Janus-like perspective in two directions, and with effect in two respects: 'outwards' toward the expression-substance and 'inwards' toward the content-substance" (Hjelmslev 1953: 58). Pure form is an abstracted structure that meets the unformed matter (purport), or as Hjelmslev formulates it, it is "projected onto it just like when an outstretched net casts its shadows on an undivided surface" (Hjelmslev and Uldall 1967: 52) and substance is in itself also "matter-less" (stofløs) as Rasmussen (1992: 361) points out.

Hjelmslev divides language into four strata, and "the reason for describing a text as consisting of four strata is the purely formal one that the components of one stratum cannot be found by analysis of components of any of the others; the strata, in other words, do not mutually conform" (Hjelmslev and Uldall 1967: 27–28). Hjelmslev's four strata are defined as content-form, content-substance, expression-form, and expression-substance. He, thus, ascribes substance and form to both content and expression and states that substance is distinguished from both matter (amorphous purport) and form. Substance in a semiotic system is the result of form structuring or "lay[ing] down its boundaries into" (Hjelmslev 1953: 52) the unformed matter. The unformed matter is not semiotic and, therefore, not part of language; it falls outside of the area of linguistic (and semiotic) investigation. This is indirectly revisited by Kress and van Leeuwen as we shall see below regarding their distribution "stratum." The content's unformed matter is the "amorphous thought-mass" (Hjelmslev 1953: 52) that is formed by the language; the expression's unformed matter is the unformed physical potential of the involved resources, for instance the abilities of the human vocal apparatus, body language, facial expression, or the affordances of the graphic tools involved.

According to Hjelmslev, there is no direct coincidence between the physical features of the sound and the categories we use to describe in phonetics. The physical sound differences are continua whereas the descriptive categories are dichotomous (Hjelmslev 1997: 107–10). We can argue in relation to many multimodal phenomena including language that an analogue descriptive approach will be productive (Boeriis 2009)—it often seems to be more productive to use systemic continua with named poles or steps/areas to which the choices adhere in different degrees, not just for expression systems such as phonetics or phonology, but also for the meaning systems.

To Hjelmslev, the units of language are neither sounds nor meanings but, rather, the "relata" between form, substance, and matter they constitute. The elements of language form are of "algebraic" nature and as such they do not have a nature-given denomination but can be named in multiple ways (Hjelmslev 1966: 94). Following Saussure's terminology and using the terms "form" and "substance," Hjelmslev argues that the manifestation is selections in hierarchies where the form is the constant and the substance is the variable (Hjelmslev 1966: 94). The theory of language structures is independent of the naming of substance. The main task for the theorist is to: (a) define the structural principles of language; (b) following from this, deduce a typology that categorizes the single language types; and (c) identify choices not (yet) manifested in actual language use by carefully foreseeing every possible choice. Thus, structure types are not necessarily manifested but are "manifestable," and this occurs in any kind of substance (Hjelmslev 1966: 94). Hjelmslev states that substance is not a necessary prerequisite for the language form; however, the language form is a necessary prerequisite for the substance. The substance is dependent on form, whereas the opposite is not the case. While this might seem reasonable, it could

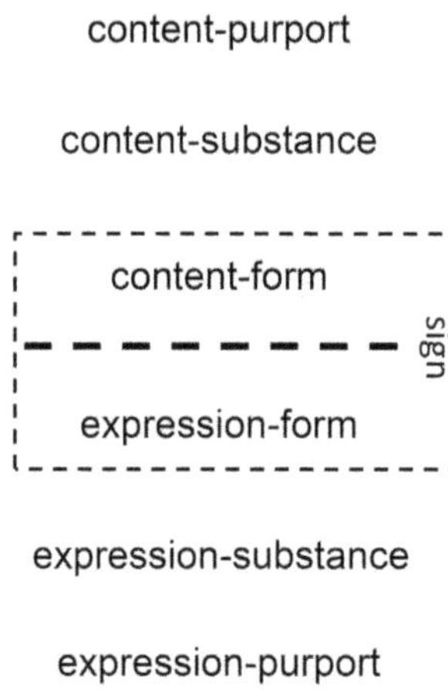

Figure 4.1 Classic modeling of Hjelmslev's stratification (see for example Nöth 1990: 67).

also be argued that this is not the whole story when it comes to the multimodal understanding of semiosis: this article wishes to state that, from a multimodal perspective, substance and form are, at least to a certain degree, interdependent in more complex ways than Hjelmslev argues. This will be elaborated below.

A widely used model of Hjelmslev's strata is similar to the visualization in Figure 4.1. Note how the expression and content only meet at the border between the two form strata.

We shall now turn to Halliday's modeling of stratification. Halliday's notion of the interfacing function between the strata of semantics and phonetics, and the social and bodily context, respectively, which is elaborated below (Halliday and Matthiessen 2004: 24–26), is not completely similar but related to Hjelmslev's modeling: it has to do with strata used for interfacing with matter, be that matter as (social) experience in the case of semantics or physical matter in the case of phonetics. Hjelmslev describes the substance strata as the result of the form strata interacting with or imposing on the unformed matter (purport) of the content and of the expression, respectively—the form imposed on matter makes substance (Hjelmslev 1953: 92).

3 Halliday's stratification of language

The Hallidayan stratification is inspired by Lamb, who sees language as consisting of three strata, namely the stratum related to meaning, called "semology," the speech-related stratum, called "phonology," and the stratum of "grammar" between the two (Lamb 1966: 1). Halliday conceptualizes meaning making as a co-articulation of phonology (and phonetics), lexico-grammar, and semantics, where each level realizes the next. Lamb gives equal priority to syntagmatic and paradigmatic relationships in the language system, and Halliday takes this one step further and gives priority to the paradigmatic axis, stating that each stratum is a network of paradigmatic relationships as a set of alternative systemic choices (Halliday 1978: 40). He stresses that each of these systems—semantics, grammar, and phonology—is a system of potential, a range of alternative choice options, and elaborates: "I would use the term network for all levels, in fact: semantic network, grammatical network, phonological network. It refers simply to a representation of the potential at that level" (Halliday 1978: 40).

Backtracking to Saussure, Halliday mentions Hjelmslev as a major inspiration for his understanding of the stratification axis (Halliday 1961, 1978: 39–40). In different stages of Halliday's systemic functional grammar, the stratification has been modeled differently. Both in very early iterations (see Halliday 1961) and in more recent iterations of the modeling of stratification, Halliday expands Lamb's tri-stratal approach to four strata and gives full elaboration to the expression side as divided into two separate strata like the two content strata (Halliday and Matthiessen 2004: 24–26). This model of a fourfold stratification (embedded in context) is, as Halliday also states, comparable though not identical to Hjelmslev's four strata as we shall see below. Using speech as an example, Halliday argues how the content side in the traditional sign is expanded into two, namely lexico-grammar and semantics, and similarly the expression side is expanded into two, namely phonology and phonetics. Thus, the relationship between the strata is described as realization where semantics realizes the relation between environment and meaning, lexico-grammar realizes the meaning in wording, phonology realizes the wording in composing, and phonetics realizes the composing in sounding (Halliday and Matthiessen 2004: 26). In a closer connection to Hjelmslev, the four strata allow a separation of the organizing function from the interfacing function. Semantics is the interface "via receptors" (Halliday and Matthiessen 2014: 25) to the social world around language, ascribing meaning to experiences and relationships seen as a connection between environment and meaning, and lexico-grammar is the internal organizing of meaning into wording. Similarly, phonetics is the function of interfacing "via motors" (Halliday and Matthiessen 2014: 25) to the bodily affordances as a biological system of resources—the sounds we are able to create as human beings—whereas phonology is the organizing function of language, that is, organizing expression into functional categories.

Realization is a one-way street in this conceptualization going upwards, and there is no relationship between strata other than between immediately adjacent ones. Hallidayan stratification is the norm in the majority of accounts of multimodal stratification. It is typically modeled either as co-tangential circles or more simply in a similar way as in Figure 4.2. As discussed by Taverniers (2011: 1102), this one-dimensional understanding of stratification prevails in what she names stage II in systemic functional grammar, whereas in stage III Halliday moves on to interpret stratification as "metaredundancy circles" (Halliday 1987). This approach is inspired by the introduction of dynamic open systems by Lemke (1984). Here, the relationship between strata is to a great extend based on probability, meaning that one element on one stratum does not directly correspond with elements at another but redounds with them in the sense that the relation is more probable, typical, and predictable. In Lemke's words (1995: 168), "two things are 'redundant' when they go together in a predictable way." And this entails that in a given system "two levels . . . are in a redundancy relationship when options in one level tend to co-occur in a predictable way with options in another level" (Taverniers 2011: 1110). Stage II accounts of stratification still prevail in recent work in systemic functional

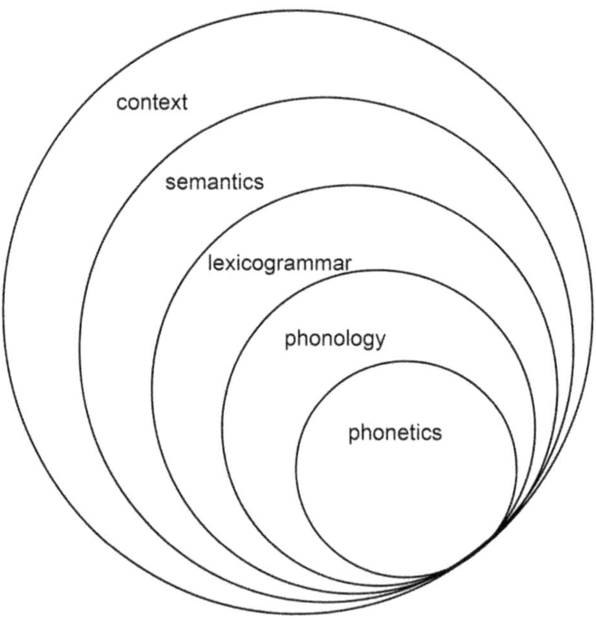

Figure 4.2 Typical model of Halliday's stratification, adopted from Halliday and Matthiessen (2004: 26).

grammar. Even though language is described as "a series of redundancies by which we link our ecosocial environment to nonrandom disturbances in the air (soundwaves)" (Halliday and Matthiessen 2014: 25) in the chapter on stratification in the recent version of *An Introduction to Functional Grammar*, the notion of "realize" is preferred to the term "redound," and the chapter still presents a rather linear conceptualization of stratification. As we shall see below, the understanding of how strata are redounding (and meta-redounding) is relevant in a multimodal conceptualizing of stratification as it does not have the same issues of directionality and linearity.

An alternative social semiotic understanding of stratification that has held a prominent position is that of Martin. He sees discourse-semantic structures as structures above the clause (Martin 1992: 20) which is similar to Hjelmslev's connotative semiotic system (Tavernier 2011: 1002). Martin's approach has had a great influence in many contexts—not least within learning and education theory. Martin's stratification also plays an important role in what might be called the Bremen school of multimodality, particularly when dealing with moving images (see Bateman and Schmidt 2012; Tseng 2014; Wildfeuer 2014). However, for the purpose of this article, Martin's understanding of (discourse) semantics as larger elements of text is not as fruitful as an understanding where semantics is, in the Hallidayan sense, the meaning part of the content, or as Hjelmslev would see it, the abstract content structures between grammar and unformed meaning-matter (see below). The relationship between the clause and larger text elements, or indeed between any conceivable kind of figure and larger units in any semiotic system, is better addressed in terms of rank or scalar levels, as smaller units within larger units but still as part of the same stratum (see Lemke (1991, 1995) on text scale; Baldry and Thibault (2005) on multimodal scale; Boeriis (2009, 2012) and Boeriis and Holsanova (2012) on dynamic rank and analytic zoom).

A very interesting point made by Halliday about "foregrounding" opens the topic of nonlinguistic systems that bypass grammar in meaning making, which shall bring us on to a discussion of a more complex understanding of the stratification relation in semiotic systems.

> A great deal of stylistic foregrounding depends on *an analogous process, by which some aspect of the underlying meaning is represented linguistically at more than one level*: not only through the semantics of the text—the ideational and interpersonal meanings, as embodied in the content and in the writer's choice of his role—but *also by direct reflection on the lexicogrammar or the phonology.*

> A text, as well as being realised in the lower levels of the linguistic system, lexicogrammatical and phonological, is also itself the realisation of higher-level semiotic structures with their own modes of interpretation, literary, sociological, psychoanalytic, and so on. These higher-level structures may be expressed not only by the semantics of the text but also by patterning at these lower levels; when such lower-level patterning is significant at some higher level it becomes what is known as "foregrounded." *Such foregrounded patterns in lexicogrammar or phonology may be characteristic of a part of a whole text, or even of a whole class or genre of texts*, a classic example being the rhyme schemes of the Petrarchan and Shakespearean sonnets as expressions of two very different modes of artistic semiotic (patterns of meaning used as art forms). (Halliday 1978: 138–39; emphases added)

This relationship between, to put it simply, semantic meaning and its direct phonetic or phonological realization, skipping the lexico-grammar stratum, is part of the very motivation for proposing a spatial remodeling of the stratification structure. This kind of meaning making is not at all exotic or rare in multimodal communication, and the article proposes that it is a fundamental part of meaning making in any constellation of semiotic resources.

Next we shall move on to how the concept of stratification has been utilized within multimodal theory, where it has not played a very central role, probably due to definitional problems concerning mode and multimodality and unclear use of concepts like semiotic systems.

4 Kress and van Leeuwen's multimodal stratification

Stratification has been a topic of some discussions in multimodal theory (see Lemke 1991, 1998, 2000; Lim-Fei 2004; Baldry and Thibault 2005; Matthiessen 2007; O'Halloran 2008; Boeriis 2009; Bateman and Schmidt 2012) but has not held as prominent a position as it has in systemic functional linguistics. However, Kress and van Leeuwen's alternative approach to stratification has had some impact in the multimodal field with their "four domains of practice in which meaning is often made" (Kress and van Leeuwen 2001: 4). Kress and van Leeuwen explicitly refer to Halliday by calling their domains "strata" to "show a relationship to Hallidayan functional linguistics, for reasons of the potential compatibility of description of different modes" (Kress and van Leeuwen 2001: 4). They coin the four "strata": discourse, design, production, and distribution. They exemplify this clearly in the following passage: "Speakers need access to discourses, knowledges which are socially structured for the purpose at hand; they need to know how to formulate these knowledges in the appropriate register and how to embed them in an (inter)active event; and they need to be able to speak" (Kress and van Leeuwen 2001: 9).

In this model, "discourse" is described as "socially constructed knowledges of (some aspects of) reality" (Kress and van Leeuwen 2001: 4), and discursive practice as "the ability to select the discourses which are to be 'in play' on a particular occasion, in a particular 'text'" (Kress and van Leeuwen 2001: 30). This notion is comparable to Hallidayan semantics in that it, of course, has to do with the meaning part of the content side of the sign. It also depends on the surrounding social environment to provide the culturally available discourses. (which is, in fact, not entirely unlike Hjelmslev's content-substance as the result of form meeting matter [content purport], where matter is not too far from a social repertoire of yet unformed but available meanings.)

To Kress and van Leeuwen, "design" "stands midway between content and expression. It is the conceptual side of expression and the expression side of conception" (Kress and van Leeuwen 2001: 5). As such, this definition is to a certain extent quite comparable to Halliday's lexico-grammar as the stratum between semantics and "expression." Design is both the formulation of a discourse, a discourse interaction, and a combination of modes (Kress and Jewitt 2003: 21), and it refers to how people plan to utilize available resources at particular communicative events in time and environment in order to achieve their communicative purpose as text makers (Kress and Jewitt 2003: 17). The planning aspect is different from Halliday's lexico-grammar as Kress and van Leeuwen's design stratum (or even praxis) is, in a way, the strategic planning of the semiotic event, the decision about which resources to realize in the text and how, before they are actually implemented. It involves the evaluation of the way the most apt resources are chosen for the communicative purpose. (For further discussion about strategies for semiotic aptness, see Kress 2010.) The semiotic modes drawn upon in the design remain abstract here, but Kress and van Leeuwen expand upon the notion of "intermediate productions" such as music scores or blueprints (Kress and van Leeuwen 2001: 21). Even though these "production plans" are interesting as part of a highly conventionalized meaning-making process, they are not regarded as part of the design in this article. They represent the semiotic production and function as aids for planning and maintain the result of the abstract design process for a later production, and as such they are full texts in themselves, not an integrated part of the meaning making in the text they support.

The physical expression strata in Kress and van Leeuwen's (2001: 7) modeling differ somewhat from both Halliday, Hjelmslev, and most traditional linguists, in that they argue that the expression plane does not merely realize the content, but it adds further meaning in itself. Expression matter matters to Kress and van Leeuwen (2001: 69), and it plays a pivotal role in meaning making. "Production" is the materialization of the choices made into perceivable material—the abstract made concrete—it is "the articulation in material form of semiotic products or events" (Kress and van Leeuwen 2001: 21). The authors provide examples, such as elaborate discussions of meaning making involving voice qualities (Kress and van Leeuwen 2001: 82–85). Production is not a simple execution of the strategic plan, since, regardless of how inconspicuous the production may be, it always adds further meaning. The Kress and van Leeuwen's understanding sees "materiality as a source of difference, and hence of meaning" (2001: 70), and this we shall take into consideration below.

"Distribution" is about the relationship with the physical—(the vehicle of) "the technical 're-coding' of semiotic products and events" (Kress and van Leeuwen 2001: 21). Kress and van Leeuwen crucially state that meaning is made at every level, and the choice of distribution, such as medium and materiality, is in itself of semiotic significance. They use music as an example of how the distribution may affect the meaning making directly. Distribution is clearly not similar to Hjelmslev's expression-substance; it is perhaps rather more related to the expression unformed matter (purport)—the media involved (and how choosing media in itself is meaning making). In this notion of distribution, we see a fairly obvious focus on the bodily abilities to create the appropriate sounds—the physical affordances—which has more resemblance to the Hallidayan phonetics than might appear at first glance. It is, in a way, the (ability to) interface with the body motors. The major difference is, of course, that Kress and van Leeuwen are not talking about the system's ability but rather the communicator's ability to utilize the apt physical (bodily) resources.

Kress and van Leeuwen's stratification model describes the process of "a text in becoming," and this is a fairly linear production-oriented temporal progression from top to bottom. Kress and van Leeuwen argue convincingly that meaning making takes place at every level. The main difference is obviously, as already discussed, that Kress and van Leeuwen's strata are seen from within the domain of the sender in a typical communication model, whereas Hjelmslev takes the perspective from inside the semiotic system, as does the Hallidayan approach that sits somewhere in the middle of the two approaches with the addition of a crucial social dimension.

The phenomenon Kress and van Leeuwen (2001: 74–5) term "experiential meaning potential" is discussed under "production." It is the bodily experiences of making the same or similar physical action with the perceiver's own body, enabling the perceiver to perceive and distinguish differences that are not (yet) elaborately formed as meaningful because of the recognized physical actions involved. Not every aspect of the meaning making is formed. For instance the use of the voice tension to convey "tension-ness" (van Leeuwen 2005: 33) or in the strokes of a pencil on paper surface the "handmade-ness" can be deduced from the tiny irregularities of the line (Johannessen 2010: 102–13). As a mode matures and expression resources sediment, this may cause forms to develop, and the choice becomes a (more deliberate) part of the system with both a form and a substance.

Similarly, Kress and van Leeuwen's concept "provenance" (2001: 23) is the idea that signs may be imported from one context into another to fix meaning where a specialized conventionalized semiotic system has not yet evolved. It is not completely clear why this is described under the production stratum and not as an aspect of discourse, but it might have to do with Kress and van Leeuwen's focus on the multilayered text production process. This clearly shows some of the incompatibilities between the stratification of Kress and van Leeuwen and Halliday (and Hjelmslev), but there are some similarities as well. Regarding provenance and experiential meaning potential, Kress and van Leeuwen (2001: 79) say that "they are the materials of which modes are made." In that sense, provenance is (akin to) importing (yet) unformed content matter into the system and thereby invoking a resonance with other similar experiences of meaning, so the meaning becomes associative ("-ness" meanings, like Barthes's 1977 famous "Italianess" in a Panzani ad). This is quite comparable to Hjelmslev's concept of secondary semiotics, specifically the connotative semiotics (Taverniers 2011: 1102). Similarly, Kress and van Leeuwen's (2001: 22) concept of experiential meaning potential means importing unformed expression. This article argues that these two kinds of "formless meaning making" are crucial parts of the process of meaning making, resulting in an understanding of a non-hermetic system where complexities are imported into the system to create (creative) interesting meanings. Then, as the same choice might become more common, it will be conventionalized into the system and, as such, formalized as well.

Even if Kress and van Leeuwen's stratification is clearly quite different from Hjelmslev's and Halliday's and has a production-oriented approach, we can identify some insights provided by it that are worth considering. As we return to remodeling multimodal stratification below, these insights about multimodal meaning making will be part of the scaffolding.

5 On multimodality

Before moving to the reorganization of the modeling of stratification, we will briefly turn the attention to multimodal text and discourse. Lemke distinguishes between an "object text," which is the text as physical change in the world, and a "meaning text," which is text as semiotic change in the world (Andersen et al. 2015: 122). Kress focuses more on the latter and sees text as the result of a communication process and as such that can be "any semiotic entity, which is internally coherent and framed, so that I can see this entity as separate from other entities" (Andersen et al. 2015: 82). Martin (Andersen et al. 2015: 49) and Matthiessen (Andersen et al. 2015: 20–1) focus on text as a matter of instantiation of resources from a semiotic system—the text is the instantiated choices, so to speak. On the very first page of *Multimodal Discourse: The Modes and Media of Contemporary Communication*, Kress and van Leeuwen (2001: 1) use the term "Gesamtkunstwerk" to describe the interplay of resources in multimodal meaning making. Matthiessen (2007: 3) clarifies this as follows: "All multimodal presentations unfolding in context are like Richard Wagner's conception of an opera as a Gesamtkunstwerk, where the different contributions are woven together into one unified performance." In this article, the weaving Matthiessen speaks of is understood as fully multimodal. A widely held understanding, which we can term polymodality, sees the resources as confined to predefined modes, which are then brought into text as whole systems. However, this understanding has been increasingly contested, starting with Kress and van Leeuwen (2001: 124) who doubt as to "whether grammars of distinct modes are quite so uncontentiously 'there' as our own efforts in relation to images, for instance, suggest." Baldry and Thibault (2005: 18) state that the separation of resources into modes is a mere analytical abstraction as "different resources are analytically, but not constitutively, separable in actual texts," and they continue that this abstraction may show both generic (i.e., standardized) and text-specific (i.e., individual, even innovative) aspects.

Following this, the position in this article is that the semiotic resources are not *a priori* assigned to any modes (at least not any of the traditional conceptions of these) but, rather, instantiated into the text to fulfill certain roles in the overall multimodal ensemble. (For a further discussion of polymodality versus multimodality, see Boeriis 2008, 2009.) This approach stems from the functional understanding of language inherited from Halliday. The descriptive aim of the theory is not to make "grammar as rules," but rather to make "grammar as frequent (combinations of) choices." In other words, the grammar does not describe rules but social praxis. This has been discussed and problematized, for instance when van Leeuwen asks, "Have we created a monster" (Andersen et al. 2015: 110) in relation to the question about whether grammar is also imposing itself on behavior, creating intervention, regulation, and rules. This is an important objection, but this article wishes to maintain the functional descriptive objective of social semiotic theory: the aim is to describe conventions as patterns of typical systemic choices. The description of separate modes and their interplay in texts has been the norm in much of multimodal theory so far; but, as described elsewhere (Boeriis 2009: 67–70), this is an unsatisfactory approach. Based on the base assumptions in social semiotic multimodal theory, the logical reasoning here is simple: if the purpose of the social semiotic approach is to describe patterns as seen in instantiated text and "texts of all kinds are *always* multimodal, making use of, and combining, the resources of diverse semiotic systems" (Baldry and Thibault 2005: 19, original emphasis), then the description should be of those occurrences. That would imply some revisions of our common analytical abstractions and, however complex and intertwined this may become, our descriptive categories have to be derived from the semiotic phenomena we do, in fact, encounter. Analytical abstractions are not necessarily problematic in themselves, but they become problematic when the abstractions are treated as ontologically real—when language, for instance, is conceptualized as a thing. This discussion will not be elaborated further here, but the radical multimodal approach is pivotal to the understanding of multimodality in this article.

6 Multimodal stratification in hemispheres

The following proposed remodeling of multimodal stratification takes inspiration from all of the three approaches above, as it tries to remedy some of the issues mentioned in an attempt to optimize the way stratification is conceptualized graphically.

Even though Hjelmslev does take into account resources other than the strictly linguistic, such as gesture and facial expression, he does not present a multimodal approach as such. As mentioned above, Hjelmslev sees the sign relation strictly as a matter of the relationship between the two form strata. Following Gibson (1986), semiotic artifacts are only a part of the semiotic array, but in this article the meaning making, which is in focus, is exactly the social meaning making. This is derived from a social semiotic understanding of social communicational meaning making, where artifacts are seen as texts and where the texts are the material available for analysis (Baldry and Thibault 2005: 192; Boeriis 2009: 98–102). The focus of this article is the communicative intentional meaning making, where the purpose is to convey meaning to somebody by means of semiotic artifacts—namely the text. This is the main area of interest to Halliday and to Kress and van Leeuwen as well. Even though for the latter the focus on materiality is more pronounced, it is still social semiotic meaning-making. Affordance is a much-discussed term in multimodal theory (see Andersen et al. 2015)—it has been introduced into social semiotics from the Gibsonian conception (among others by Gibson 1986; Lemke 1995; Baldry and Thibault 2005; Kress 2010). The concept of affordance plays an important role in the following, not in the sense of the potential for interaction between an animal and elements in its ecological surroundings but, rather, as a derived abstract understanding of different kinds of semiotic potentials in the semiotic system available in the process of meaning making.

The model presented in this article places the four strata not in a hierarchy (unlike the typical model of Hallidayan stratification seen above) but, rather, as a matrix with the form-substance distinction on one axis and the content-expression distinction on the other and as such explicitly modeling both Hjelmslevian dimensions of connection and manifestation (Taverniers 2011: 1105; see also Guattari 1996: 149 and Taverniers 2011: 1103 for similar modeling of stratification).

From this matrix, it is possible to generalize the four basic stratificational functions as hemispheres where north is the content hemisphere, south is the expression hemisphere, west is the form hemisphere, and east is the substance hemisphere. The basic functions then combine as strata in each quadrant. In Hallidayan terms, the western hemisphere is the organizing function and the eastern is the interfacing function; the northern is content and the southern is expression. The first immediate advantage of this model is that the hierarchy of the strata is avoided, as the strata are integrated more explicitly into one unit. This is stressed further by inscribing the matrix into a circle.

In order to avoid any potential confusion derived from using mode-specific terminology when discussing a multimodal axiom, we can adopt Hjelmslev's more general terminology. We take over Hjelmslev's term "cenematics" for the study of expression substance such as phonetics in speech (even if the original Greek meaning has to do with being without meaning or "empty"),

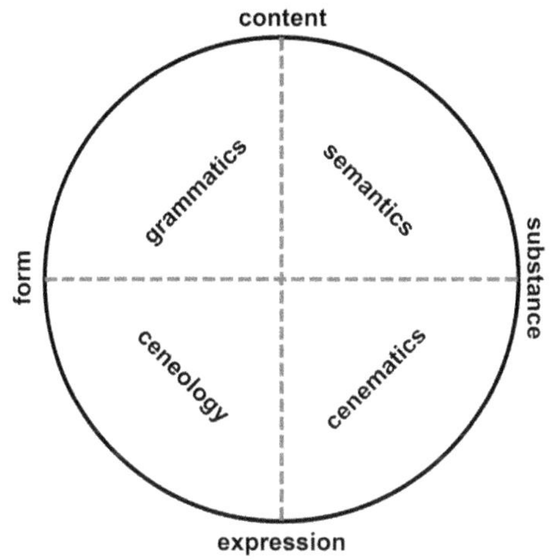

Figure 4.3 Basic model for multimodal stratification.

and in connection with that, this article proposes "ceneology" as the study of expression-form such as phonology in speech. For the content-substance, this article also proposes Hjelmslev's term "semantics," which is commonly used, and similarly for the study of content-form, this article proposes the term "grammatics" (Hjelmslev 1966: 79).

7 The multimodal context

Following Halliday's modeling of the strata as "embedded in context" (Halliday and Matthiessen 2004: 26), the proposed model also embeds all strata into the context in a way that opens new opportunities to discuss the conceptualization of context. We see how it is possible to operate with a context to each hemisphere and we discover how we may benefit from thinking of how each stratum has its own immediate context.

The context of the substance hemisphere can be seen as corresponding to the unformed matter. In the case of the expression-substance, the unformed matter is the cenematic affordances of the physical medium involved. Expression-substance is not, as Halliday says, about how phonology as stratum is an interface to the body motors, but, rather, an interface to the physical choices afforded by the chosen medium (for instance the body) in light of the available forms; the expression matter is the cenematic physical affordances of the medium involved. This could be, for instance, due to the affordances provided by the medium of the human voice, by photography, or by the pen and paper. The context of the content-substance is the unformed social matter of the social world involved, which provides the affordances of the possible meanings; it is, so to speak, the available conceptualized meaning provided by the social context.

The context of the form hemisphere is the available options for giving form afforded by the environment around the semiotic system. The context of the content-form affords the available grammatic systems of the culture in a given situation and similarly the context of the expression-form affords the available ceneologic systems.

8 Stratal relations

A tight dialectic relation of realization between the strata is problematized here, as the relationship between strata can be seen as much more complex than merely a linear one-directional connection upwards. In the multimodal stratification model in Figure 4.3, the classic linear vertical realization relationship presented by Halliday is represented clockwise from expression-substance via expression-form and content-form to content-substance. In meta-redundancy terms this is the redundancy between cenematics and ceneology seen as meta-redundant with grammatics, and this meta-redundancy is again meta-redundant with semantics. But, as discussed above, this is only part of the story. The circular modeling accounts for more relationships between strata such as content redundancy, form redundancy, expression redundancy, and substance redundancy. The model presented here describes explicitly how meaning making takes place at all strata as redundancy relationships between semantics and any of the other strata.

From a multimodal perspective, following Kress and van Leeuwen, the expression-substance stratum is an immensely rich area of the semiotic system where expression-form meets the unformed matter and in this meeting semiotic reverberations going across the content-expression line are created, which influence the process in the content-substance directly. This implies a direct redundancy relationship between cenematics and semantics, which is closely related to Kress and van Leeuwen's experiential meaning potential (see for instance Kress and van Leeuwen 2001: 10–11) where our (bodily and perhaps other) experiences of performing similar acts inform our understanding and as such provide the background for the meaning. These relations have often been left out or treated as secondary in the overall meaning. The presented model lends itself to viewing them as fully realized meaning-making relationships (redundancies) we experience but often leave out when accounting for the meaning making. Hence, the model encompasses what we have so far thought of as mere "tacit meaning," in the sense that it is meaning that we perhaps "feel" but do not use as sophisticatedly and intentionally as the linguistic. The "non-linguistic" resources (often termed extra-linguistic) have not been considered as central in the more monomodal approach to (linguistic) theories about language and communication in the humanities and have not received the same exhaustive attention as language. This is changing and, not least with the multimodal theory, we may direct our attention to all sorts of expressive phenomena, for example, sharp or blunt shapes, saturated or muted colors, hard or soft shapes, loud or quiet sounds, etc. These are all meanings that so far have gotten lost in translation when dealing with multimodal texts. It is a question of not only attending to the meaning making of new modes besides language, which is important in itself, but also dealing with the physical experience or bodily knowledge involved, when, for instance, van Leeuwen speaks of "experiential meaning potential" (Kress and van Leeuwen 2001: 22–23). The vertical redundancy between cenematics and semantics "short-circuits" the linear modeling of the flow of realization and enables discussions of the stratal role of expression in meaning making.

We have also modeled how expression form can help realize expression-substance, that is, creating meaning without grammar. This is the case when phonologic choices have direct meaning without distinguishing grammatic differences. For instance intonation contours define a question or a demand, as well as declarative questions without any grammatic intervention (speech acts as semantic phenomena). Another interesting redundancy relationship is the diagonal relationship between expression-substance and content-form where the expression-substance, so to speak, "creates grammar" directly, without the expression form of cenematic modeling, that is when grammar is not determined by distinguishing features but, rather, by graphic signs like punctuation such as question mark for interrogative or colon representing projection, or when bold or caps may mean "shouting," and quotation marks or italics may indicate projection.

As discussed above, cenematics is (the description of) the systems of available choices, which emerge as a result of form (ceneologics) meeting matter (medium). A substance can be conceived as form meeting matter (as Hjelmslev states), but we prefer to see this relationship as interdependent, so when Hjelmslev says that substance needs form, it is only part of the story, since (new) form can also be demanded in certain ways. We can think of a social semiotic need or pressure for a certain form to perform a certain function in the semiotic praxis. There can be a need to form matter to get a certain substance to function for a certain communicative purpose. This need emerges from new semiotic innovations as well as from new options afforded by technological innovation or discoveries of new interesting matter that we wish to put to semiotic use. All of these create a demand for new forms to encompass the new matter. One might think of emoticons as an example. As the written language becomes more dialogical in the digital chat and text messages, a demand arises for expressing the delicate meanings conveyed by facial expressions and body language in face-to-face interaction. Emoticons can provide these modifications of the verbal content, often in terms of interpersonal (modality) functions. Similarly, a semiotic pressure for the creation of new meaning can lead to new (technological) innovations of matter, enabling new affordances for semiotizing matter into meaning making, as is the case for many art forms doing semiotic innovation.

Now we shall move on to consider how the new modeling of stratification relates to a co-modeling of one of the other most important axes relevant for multimodality, namely instantiation.

9 Further implications—stratified instantiation

In the illustration on the cover page of *An Introduction to Functional Grammar 3rd ed.* (2004), Halliday and Matthiessen present a model combining stratification and instantiation into the same matrix (also including metafunction), providing a graphic overview of the structure of language (the model is elaborated within the book on pages 30–33). The presented intersection of the strata with the instantiation cline can also provide interesting perspectives in relation to multimodal meaning making, and the following section is a presentation of how instantiation maps onto the multimodal stratification model.

Instantiation is, in Halliday and Matthiessen's view (2004: 26), a description of the cline between the overall system and the text. This allows us to talk about the instantiated choices in light of what is not instantiated. In multimodal text creation, we do not choose from the system of a single mode; we choose from multiple system elements and their conventions of combinations.

Here we propose to implement a combined modeling of instantiation and stratification so as to be able to describe both semantic, grammatic, ceneologic, and cenematic instantiation. As mentioned above in relation to the meta-redundancy approach, the resources are not bound in a tightly welded 1:1 connection, so that, for instance, one grammatic choice realizes only one exact semantic choice. One can, for instance, choose between disparate combinations of systems in which grammatic meaning will be used to realize a semantic meaning—and this can be done via choices between very different resources in the traditional modes. This differs from what Matthiessen proposes in his discussion of multimodal stratification (Matthiessen 2007), where he sees the multimodal resources as divided in the expression strata but united in the content strata, resulting in the multimodal resource integration being a vertical stratification process. This article proposes a view where resources integrate in instantiation—which in relation to the multimodal stratification model means a radial integration of resources as will be elaborated below. One interesting consequence of this understanding of radial instantiation is that grammatical metaphor is perhaps less of a curiosum but, rather, a less common but just as correct way of realization. However, this will not be discussed any further in this article.

In a footnote, Halliday and Matthiessen talk about the two kinds of systems at play in his terminology, namely the overall system of language (and, as such, equivalent to the traditional mode) and then the (sub-)systems described in each overall systemic entry condition (Halliday and Matthiessen 2004: 26). This double use of the system as both hyper-system and subsystem is relevant for the discussion of multimodal text instantiation. As argued above, the traditional modes as hyper-systems are not sufficient for modeling contemporary multimodal communication. The modes are dissolving into what might be construed as a huge amorphous repertoire of multimodal resources from which we choose what to instantiate into each text more freely than has so far been described. This dissolution may be the first part of a reorganization of our conceptualization of semiotic resources, and perhaps the amorphous repertoire will condense into new reorganized complex hyper-systems involving the intertwining of resources we have hitherto not theorized together. Some choices of combination become so common that they form a rule in the system—a conventionalized combination—so in this view, the system consists of both choices of system and choices of combination. Part of the resources are the connectors—both intra-stratal connectors and inter-stratal connectors. We may metaphorically think of it as similar to a chemistry class molecule set—in order to form molecules (texts), both atoms (subsystems) and bonds (connectors) need to be part of the set (hyper-system). Some fit only some elements, while some fit many. Further examination of this relevant topic is unfortunately not possible in this article.

If we let instantiation happen along the diameter of the circle, so that the instance is in the center and the system is at the border, we get some interesting possibilities. Halliday argues that when we study larger corpora of text, we are able to identify registers or text types between the system and instance along the instantiation cline (Halliday and Matthiessen 2004: 27). This notion of "typical combination of choices" is extremely relevant in relation to multimodal texts since it focuses on combination regularities. As such, the registers or text types are regular patterns of typical choices occurring together—typical ensembles of instantiated choices. If we map the cline in three layers, we get an intermediate area, which is for typical constellations of resources, the text types. Thus, at one end we have the potential resources (periphery), at the other end (center) we have the selected resources in the text, and between them we have frequent resources that often appear together in the text type "zone." The model now provides for describing both ceneologic, cenematic, grammatic, and semantic clines and encompassing the system potential, the individual types, and the instantiated text as stratified unities.

The cenematic cline extends from the text through frequent choices in text types to the system of possible expression-substance choices afforded by the context of the medium. The ceneologic cline goes from the instantiated expression forms in the text through frequent form choices in text types to the system of possible choices afforded in the context of medium forms. The grammatic cline runs from the instantiated expression forms choices instantiated in the text via frequent choices in text types to the overall grammatic system afforded by the grammatic social context. We can describe the text types from a content-form perspective as registers, here meaning common grammatic choices for certain text types. The semantic instantiation cline runs from the instantiated expression-substance via frequent choices in the text types to the overall semantic system afforded by the social context.

Following Halliday's embedding of the strata into a context, which can in itself be seen as an instantiation cline between culture and situation (Halliday and Matthiessen 2004: 28), the multimodal context can also be mapped as an instantiation cline surrounding the system-instance cline. Culture would then be the outer rim at the system end, as the overall potential of the culture and

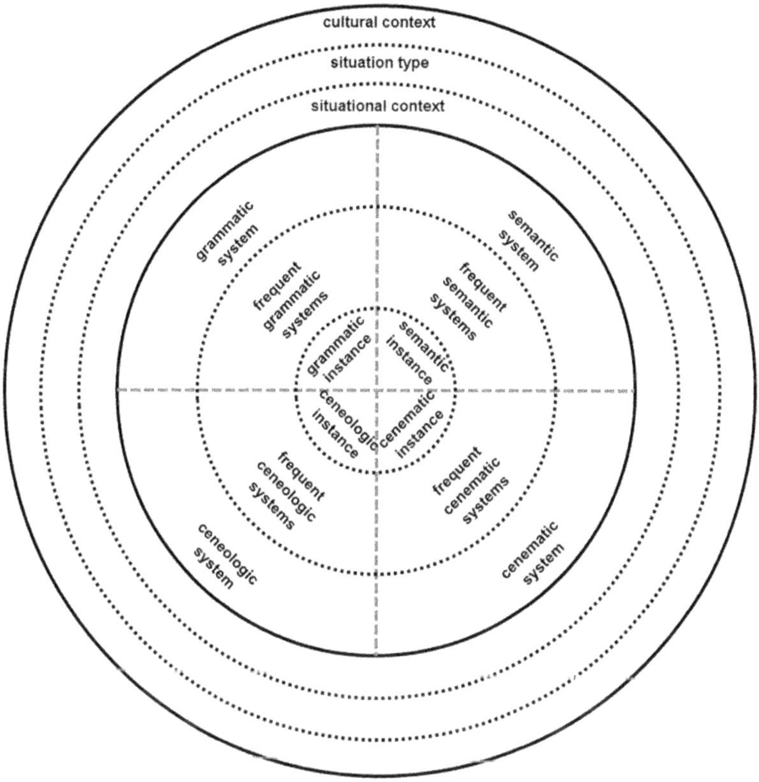

Figure 4.4 Model of multimodal stratification and instantiation.

situation would be at the inner rim as the instantiated situation. Between the two, we place the situation type (institution) as Halliday names it. The culture (or cultures) sets the scene, and the situation affords the possibilities of the environment.

If we map stratification of both the system-text relationship and the culture-situation relationship onto the system, it looks like the model given in Figure 4.4.

10 Conclusion

The article addresses the axiomatic grounds for how we conceive of multimodality, and it examines how a multimodal understanding of communication might benefit from a remodeled approach to stratification. It demonstrates how a conceptualization of multimodal stratification can draw on aspects from: (a) Kress and van Leeuwen's four articulation praxis levels; (b) Hjelmslev's division of the expression-meaning relationship into four strata; and (c) Halliday's widely used context-embedded stratification. The (stage II) Hallidayan model of stratification is implicitly hierarchical since the vertical stratification renders a fairly linear hierarchical relation between the strata, which is only reasonable if one chooses to conceptualize language as a single secluded semiotic system (an understanding that has prevailed in linguistics). Multimodal research, not least in the social semiotic branch, has focused less on discussing these basic axioms from a multimodal perspective, exploring what the multimodal understanding provides to, and demands from, the theoretical axioms. The article opens a discussion of what may have to be altered to encompass a plurality of semiotic systems. It demonstrates how the notion and modeling of stratification can be discussed in relation to a multimodal approach to semiotic modes and texts, and it argues that there is a need to revisit the modeling of stratification in light of the multimodal turn.

The remodeling makes way for considerations about the relationship between strata such as a diagonal realization relation between cenematics (expression-substance) and grammar (content-form) and, similarly, a diagonal realization relation between ceneology (expression-form) and semantics (content-substance). It also provides opportunities to map out some interesting types of contexts in relation to each stratum. This may provide interesting new ways of thinking about resources and their instantiation, about the relationship between strata and the meaning making in Kress and van Leeuwen's production and distribution strata. However, the full consequence of these new approaches to realization relations needs further investigation.

The article implicitly invites further discussions of the multimodal text and multimodal sign function. It seems relevant to rethink the sign relationship as perhaps not limited to the relation between forms but as rather more complex relations between the four strata (and the context). Similarly, the multimodal text may be understood as a local instantiation of resources—in a way a local semiotic system—instantiated by very disparate semiotic resources. This, of course, also needs to be elaborated and examined in much more detail than is possible here.

The article also demonstrates how it is possible to address experiential meaning potential and provenance in the new model. The model encompasses some of the direct un-grammaticalized meanings of the expression side, which is perceived (tacitly) but oftentimes overlooked, such as more bodily experienced meanings of signs that are brought to attention in the article by way of Kress and van Leeuwen's insistence that matter matters. In terms of provenance, it is the imported meaning, not from the (not yet developed) regularities of the semiotic system itself but from the context of similar cultural content experiences. As such, experiential meaning potential and provenance are two cases of importing unformed content-substance from the context—one from the social (content) context and the other from the physical (expression) context.

The discussion of semiotic affordance above in terms of Hjelmslev's semiotic substance sketches a new way of conceptualizing and defining affordance in multimodality as concerning the interaction between the semiotic system and its context. This differs somewhat from the Gibsonian approach, as it is, so to speak, the affordances of the form-matter relationship: When form and matter meet, certain affordances arise, and these are described in substance. Following this, it was argued how form, substance, and unformed matter are, in fact, all interdependent. "We seem to be at an odd moment in history, when frames are dissolving everywhere, and formerly clear boundaries are becoming ever more blurred. It is not therefore surprising that the same may be happening with representational resources" (Kress and van Leeuwen 2001: 125).

Our view on things is changing—phenomena that used to have little semiotic meaning attached to them (often due to the fact that they demanded specialized technical skills) have become semiotic resources in today's everyday text production environment due to the vast development in digital text production in different media. Hence, we may need to revisit the traditional modes, dissolve them, and then theorize the more general semiotic mechanisms at play, or create new notions of modes. We may view our semiotic potential as a giant generic "semiotic repertoire" of resources available as semiotic choices (tense, hue, information structure, process types, lighting, etc.). The task before us is systematization, not of specific systems, but of generic semiotic systems, mapping out the basic general potential of overall semiotic resources seen as the vast multimodal meaning potential including the conventions of picking out and combining resources.

Many of the issues raised in this chapter merit further research, and this first inquiry into how we can begin to conceptualize semiotic phenomena in a changing multimodal world may spawn fruitful follow-ups.

Acknowledgment

I wish to thank the members of Centre for Multimodal Communication at University of Southern Denmark, Denmark, for invaluable feedback and discussions on several occasions about the topics of this article. And I wish to express gratitude to the editors of this book and the external reviewer for the extensive conscientious feedback on the chapter.

References

Andersen, T., M. Boeriis, E. Maagerø, and E. Seip-Tønnessen (2015), *Social Semiotics. Key Figures*, New Directions, London: Routledge.

Baldry, A. and P. J. Thibault (2005), *Multimodal Transcription and Text Analysis*, London: Equinox.

Barthes, R. (1977), "Rhetoric of the Image," in *Image—Music—Text*, trans. S. Heath, 32–35, New York: Hill and Wang.

Bateman, J. and K. H. Schmidt (2012), *Multimodal Film Analysis. How Films Mean*, London/New York: Routledge.

Boeriis, M. (2008), "Mastering Multimodal Complexity," in N. Nørgaard (ed.), *Systemic Functional Linguistics in Use*, 219–36, Odense Working Papers in Language and Communication, Vol. 29, Odense: University of Southern Denmark.

Boeriis, M. (2009), *Multimodal Socialsemiotik og Levende Billeder* (PhD thesis), Odense: University of Southern Denmark.

Boeriis, M. (2012), "Tekstzoom. Om en dynamisk funktionel rangstruktur i visuelle tekster," in T. H. Andersen and M. Boeriis (eds.), *Nordisk Socialsemiotik. Pædagogiske, multimodale og sprogvidenskabelige landvindinger*, 131–54, Odense: Syddansk Universitetsforlag.

Boeriis, M. and J. Holsanova, J. (2012), "Tracking Visual Segmentation: Connecting Semiotic and Cognitive Perspectives," *Visual Communication*, 11 (3): 259–81.

Gibson, J. (1986), *The Ecological Approach to Visual Perception*, Hillsdale, NJ/London: Lawrence Erlbaum.

Guattari, F. (1996), "The Place of the Signifier in the Institution," in G. Genosko (ed.), *The Guattari Reader*, 148–57, Oxford: Blackwell.

Halliday, M. A. K. (1961), "Categories of the Theory of Grammar," *Word*, 17: 241–92.

Halliday, M. A. K. (1978), *Language as Social Semiotic. The Social Interpretation of Language and Meaning*, London: Edward Arnold.

Halliday, M. A. K. (1987), "Language and the Order of Nature," in N. Fabb, D. Attridge, A. Durant and C. MacCabe (eds.), *The Linguistics of Writing: Arguments between Language and Literature*, 135–54, Manchester: Manchester University Press.

Halliday, M. A. K and C. M. I. M. Matthiessen (2004), *An Introduction to Functional Grammar*, 3rd edition, London: Arnold.
Halliday, M. A. K and C. M. I. M. Matthiessen (2014), *Halliday's Introduction to Functional Grammar*, 4th edition, London/New York: Routledge.
Hjelmslev, L. (1953[1943]), *Prolegomena to a Theory of Language*, Baltimore: Indiana University Publications in Anthropology and Linguistics.
Hjelmslev, L. (1966[1943]), *Omkring Sprogteoriens Grundlæggelse*, København: Akademisk Forlag.
Hjelmslev, L. and H. J. Uldall (1967), *Outline of Glossematics. A study in the methodology of the humanities with special reference to linguistics*, 2nd edition, Copenhagen: Nordisk Sprog- og Kulturforlag.
Hjelmslev, L. (1997[1963]), *Sproget*, København: Munksgaard – Rosinante.
Johannessen, C. M. (2010), "Forensic Analysis of Graphic Trademarks. A Multimodal Social Semiotic Approach," PhD diss., Institute of Language and Communication, University of Southern Denmark, Odense.
Kress, G. and T. van Leeuwen (1996), *Reading Images. The Grammar of Visual Design*, London: Routledge.
Kress, G. and T. van Leeuwen (2001), *Multimodal Discourse. The Modes and Media of Contemporary Communication*, London: Arnold.
Kress, G. and C. Jewitt (2003), *Multimodal Literacy*, New York: Peter Lang.
Kress, G. (2010), *Multimodality. A Social Semiotic Approach to Contemporary Communication*, London: Routledge.
Lamb, S. M. (1966), *Outline of Stratificational Grammar*, Washington: Georgetown University Press.
Lemke, J. L. (1984), *Semiotics and Education*, Toronto: Victoria University.
Lemke, J. L. (1991), "Text production and dynamic text semantics," in E. Ventola (ed.), *Functional and Systemic Linguistics: Approaches and Uses*, 23–38, Berlin: Mouton de Gruyter.
Lemke, J. L. (1995), *Textual Politics: Discourse and Social Dynamics*, London: Taylor & Francis.
Lemke, J. L. (2000), "Multiplying Meaning: Visual and Verbal Semiotics in Scientific Text," in J. R. Martin and R. Veel (eds.), *Reading Science: Critical and Functional Perspectives on Discourses of Science*, 87–113, London: Routledge.
Lim-Fei, V. (2004), "Developing an Integrative Multi-semiotic Model," in K. L. O'Hallaran, (ed.), *Multimodal Discourse Analysis. Systemic Functional Perspectives*, 220–46, London: Continuum.
Martin, J. R. (1992), *English Text*, Amsterdam: John Benjamins.
Matthiessen, C. M. I. M. (2007), "The Multimodal Page: A Systemic Functional Exploration," in W. L. Bowcher and T. D. Royce (eds.), *New Directions in the Analysis of Multimodal Discourse*, 1–62, London: Lawrence Erlbaum Associates.
Nöth, W. (1990), *Handbook of Semiotics*, Bloomington: Indiana University Press.

O'Halloran, K. L. (2008), "Systemic Functional-Multimodal Discourse Analysis (SF-MDA): Constructing Ideational Meaning using Language and Visual Imagery," *Visual Communication*, 7 (4): 443–75.

O'Toole, M. (1994), *The Language of Displayed Art*, London: Leicester University Press.

Rasmussen, M. (1992), *Hjelmslevs sprogteori. Glossematikken i videnskabshistorisk, videnskabsteoretisk og erkendelsesteoretisk perspektiv*, Odense: Odense Universitetesforlag.

Taverniers, M. (2011), "The Syntax–semantics Interface in Systemic Functional Grammar: Halliday's Interpretation of the Hjelmslevian Model of Stratification," *Journal of Pragmatics*, 43 (4): 1100–126.

Tseng, C. (2014), *Cohesion in Film. Tracking Film Elements*, Basingstoke: Palgrave MacMillan.

Van Leeuwen, T. (2005), *Introducing Social Semiotics*, London: Routledge

Wildfeuer, J. (2014), *Film Discourse Interpretation. Towards a New Paradigm for Multimodal Film Analysis*, London: Routledge.

5

Understanding Multimodal Meaning Making: Theories of Multimodality in the Light of Reception Studies

Hans-Jürgen Bucher

1 Introduction: A recipient's perspective on multimodality

Since Plato's dialogue "Cratylus," the question of meaning has always been at the center of any theory of language and communication (Ogden and Richards 1923; Schiffer 1972). However, a multimodal perspective on communication complicates this question in several respects. On the one hand, a theory of multimodality has to explain the meaning of not only linguistic signs but also visuals, gestures, colors, design, sounds, etc. On the other hand, meaning in communication becomes a complex and multilayered entity, raising the question of how the meaning of an orchestration of different modes is composed by individual modes. In most theories of multimodality, meaning making is approached—implicitly or explicitly—from the perspective of a speaker, a producer, or an author. This article puts the recipients at its center, asking the question: How do recipients integrate the different modalities like text, picture, sound, and design into a coherent meaning? To explain how and why reception studies can help deploy a theory of multimodality, one can go back to a technique long since applied in ordinary language philosophy by philosophers like John L. Austin (1962), William Alston (1964), or Paul Grice (1968, 1975), and which can be labeled "dialogical": one can reconstruct the meaning of a verbal expression by analyzing the replies and reactions to that expression. For example, in his article of 1956 with the significant title "A plea for excuses," Austin demonstrated how one can analyze the action of accusing someone or concepts like responsibility, freedom, free will, and justification by analyzing the *excuses* as reactions to accusations (Austin 1956/1957; see also Fritz 2005). Following

Ludwig Wittgenstein (1977, especially: 43, 77, 136; see also Wittgenstein 1980), one could say: describing the meaning of a verbal expression means analyzing the language game in which the expression can be used to make a move.

One can transfer this dialogical methodology to the analysis of multimodality in a kind of experimental way. In a reception laboratory, the reactions of recipients to different multimodal stimuli like newspapers, magazines, online media, videos, video games, or scientific presentations are documented with different combinations of methods: eye tracking, thinking aloud during and after the reception process, re-narrations, knowledge tests, interviews, or questionnaires. These experimentally elicited reactions are normally used as data to infer media effects. In this article, these data are used methodologically as communication feedbacks to the different stimuli, which allow reconstructing the meaning-making process of the recipient. Hence, the documented reactions of the participants indicate how they perceive the multimodal ensemble, what they select as relevant, which modes they combine in which order, and so on. Experimental methods, so to speak, allow reconstructing the interaction between a recipient and a multimodal stimulus. They help, on the one hand, understand the process of multimodal meaning making and, on the other, reconstruct the structure of multimodal discourse, which is reflected in the layers and aspects of the reception results.

Seen from this perspective, experimental reception studies offer a strategy to work on the two basic problems any theory of multimodality has to confront. First, the *problem of compositionality*: What, specifically, does each of the individual modes contribute to the overall meaning of a discourse, and how do they interact? The problem of compositionality lies at the center of previous multimodality research. It is labeled with the concepts of "intersemiosis" (O'Halloran 2008: 470), "intersemiotic texture" (Liu and O'Halloran 2009: 367), "semantic multiplication" (Fei 2004: 239), or "modal interrelation" (Kress 2010: 165). Behind all these concepts lies the common assumption that the whole of a multimodal ensemble is more than the sum of its modal parts (Lemke 1998). Second comes the *problem of reception*, which is more or less the mirror image of the first: How do recipients integrate the different modes and acquire a coherent understanding of the multimodal discourse?

According to these two problems, a theory of multimodality requires two components. On the one hand, one needs a theory of meaning to explain compositionality, and on the other hand a theory of communication to explain how one can promote mutual understanding by using different modes or sign systems. In contrast to sign-centered semiotics most commonly used in previous work on multimodality, this article proposes an action-oriented theory of meaning rooted in speech-act theory, linguistic pragmatics, and the Wittgensteinian concept of language games. The theory of communication to which this article refers is based on an interactional approach to communication, which integrates concepts like intention, mutual knowledge, selection, attention, and affordance (Bucher and Schumacher 2006; Bucher and Niemann 2012).

2 Eye tracking: A window to multimodal meaning making

As it is impossible to get direct access to cognition, eye-tracking data can serve as a kind of window to reception processes (Yarbus 1967; Salvucci 1999; Holsanova 2008, 2014; Holmqvist et al. 2011; Boeriis and Holsanova 2012). It indicates to which elements of a stimulus attention is allocated and which are ignored, and also in which sequential order stimuli are perceived. The dwell time on a particular part of the stimulus indicates the recipients' level of attention and interest. Eye-tracking data also reveal the quality of reception: one can deduce from the pattern of the eye movement if the visual or the textual element is read in depth or just cursorily scanned (Bucher and Schumacher 2006; Duchowski 2007). Compared to traditional retrospective methods like interviews or questionnaires, eye-tracking data are more reliable, as they are obtained directly during the reception process and, thus, provide immediate insight into the interaction between the stimulus and the recipient. However, knowing at which object someone is looking does not mean knowing what he or she actually sees. In the presented studies, therefore, eye tracking is combined with several verbal methods like re-narration, knowledge tests, or interviews in order to gain additional information from the recipients. Data from those triangulations of methods provide manifold traces of visual attention and, thus, indicate how recipients handle the problem of compositionality. Therefore, these data allow us to identify some essential requirements on a theory of multimodality, which is demonstrated with the following examples from different eye-tracking studies.

2.1 Example 1: Text-image relations

The first example is a front page of the German newspaper *die tageszeitung*, whose lead story covers Barack Obama's victory in the primaries in 2008 and his nomination as candidate for the presidency from the Democratic Party. The lead story carries the headline "Onkel Baracks Hütte" ("Uncle Barack's Cabin") and contains a photograph of the White House in Washington, DC (see Figure 5.1, p. 95).

It is obvious that the meaning of the multimodal arrangement composed of headline text, picture, and lead text is more than the sum of the individual elements. Neither the picture alone nor the headline alone has enough semantic potential to uncover the full meaning of this front-page story. And even if all three elements are taken into account, getting the full meaning is not at all self-evident: The front page in Figure 5.1 was presented to sixteen students, of whom not even half could give an adequate interpretation. On account of lack of prior knowledge, they did not understand the allusion to the famous novel *Uncle Tom's Cabin* by Harriett Beecher Stowe, nor could they identify the building as the White House in Washington, DC. Therefore, they missed the point of this image-text combination.

The eye-tracking data documented in Figure 5.2 mirror the strategy of making sense with the help of several modes. The different elements—headline, lead text, and picture—are not perceived sequentially one after the other but *alternately* and *repeatedly*: The recipient attempts to get an overall meaning by taking each element as a context for the others. Via this process of alternately interpreting the elements, they finally achieve an overall meaning or what the recipient considers as such.

The *zigzag line* of eye movements between different semiotic elements is quite typical of interpreting multimodal stimuli and has been discovered in eye-tracking studies of all different kinds—in reading newspapers as well as in using websites or watching TV, and even in real-world scenarios like visiting a museum or following a scientific presentation (see Bucher and Schumacher 2006; Holsanova, Rahm and Holmqvist 2006; Holsanova, Holmqvist and Holmber, 2008; Bucher and Schumacher 2012; Bucher and Niemann 2012). So, *recursive* reception and an *interactive* acquirement of the different modal elements seem to be a characteristic pattern of understanding multimodal discourses.

This example allows us to infer some requirements for a theory of multimodality:

1. The theory should allow us to describe multimodal discourse as a *dynamic phenomenon*: multimodal meaning making is not a one-step action; understanding multimodal discourse does not happen instantaneously but step by step and recursively. Therefore, the theoretical frame of multimodal understanding should not be a theory of media effects but a theory of interactive reception (Bucher and Schumacher 2012). According to this

Überraschung: Der **Maler Anselm Kiefer** erhält den **Friedenspreis** des Buchhandels **Seite 4**

Heute **Finale** bei **„Germany's Next Topmodel"**: Warum ist **Heidi Klums Show** so **erfolgreich? Seite 13**

NR. 8597
23. WOCHE
30. JAHRGANG
AUSGABE BERLIN
€ 1.70 AUSLAND
€ 1.00 DEUTSCHLAND
DONNERSTAG, 5. JUNI 2008

die tageszeitung

Onkel Baracks Hütte

Barack Obama schreibt Geschichte: Nie zuvor hatte ein schwarzer Politiker die Chance, Präsident der USA zu werden. Obama sicherte sich die Kandidatur bei den Demokraten – im Duell mit Hillary Clinton. Jetzt attackiert er den Republikaner John McCain im Kampf um das Weiße Haus SEITE 2, 3

Das Weiße House in Washington: Wird Barack Obama der erste schwarze Präsident, der dort einzieht? FOTO [M] KRUPPA/CARO

BKA-GESETZ

BKA-Chef Jörg Ziercke will die heimliche Onlinedurchsuchung noch häufiger einsetzen, als von der Politik vorgesehen. Warum denn das? Das Interview SEITE 5

MEDWEDJEW

Russlands Präsident Dmitri Medwedjew besucht Berlin – ganz im Interesse der Wirtschaft. Denn deutsche Firmen sind in Russland dick im Geschäft SEITE 6

FIDEL CASTRA

Der Kampf zwischen Alice Schwarzer und den jüngeren Feministinnen wird immer härter. Dabei blockiert die konservative *Emma* den Fortschritt SEITE 12, 14

BERLIN

Angstfalle U-Bahn: Mithilfe einer ausrangierten Straßenbahn versucht ein Psychologe im Wedding, Verkehrsphobiker zu therapieren SEITE 23

taz muss sein

Die tageszeitung wird ermöglicht durch 7.953 GenossInnen, die in die Pressefreiheit investieren.

Infos unter geno@taz.de oder Tel: 030 - 25 90 22 13

Aboservice: 030 - 25 902-590
fax: 25 902-680 / abomail@taz.de

Anzeigen: 030-25 90 22 -38 / -90
fax: 030 -251 06 94
anzeigen@taz.de
Kleinanzeigen: 030-25 90 22 22

Redaktion: 030 - 25 902-0
fax: 030 - 251 51 30/ briefe@taz.de

taz, die tageszeitung,
Postfach 610229, 10923 Berlin

taz im Internet: **www.taz.de**

Obamas Trouble mit Hillary

KOMMENTAR VON BERND PICKERT

Die US-Vorwahlen sind vorbei, Barack Obama steht als Kandidat der Demokraten fest. Jetzt kann er sich dem Duell mit seinem republikanischen Konkurrenten John McCain stellen. Fast. Denn dem ersten schwarzen Präsidentschaftskandidaten in der Geschichte der USA steht noch immer jene Frau im Weg, die so hart dafür gearbeitet hat, die erste weibliche Präsidentin der Vereinigten Staaten zu werden.

Obamas erste und wichtigste Aufgabe wird es nun sein, Hillary Clinton einzuhegen und ihre Fans auf seine Seite zu ziehen. Hillary Clinton dafür die Vizepräsidentschaftskandidatur anzubieten, scheint nahezuliegen. Es wäre aber der falsche Weg.

Zum einen würde nach diesen harten Vorwahlen niemand mehr glauben, dass dies eine freiwillige Entscheidung wäre. Wer aber als starke und durchsetzungsfähige Führungskraft Wahlen gewinnen will, darf nicht den Eindruck erwecken, jetzt schon zu Kompromissen gezwungen worden zu sein. Zweitens hat Clinton selbst bislang nicht einmal die Mindestanforderungen für eine solche Zusammenarbeit erfüllt. Statt am Dienstagabend die Niederlage einzugestehen, Obama in den Himmel zu loben und zu versprechen, ab jetzt mit all ihrer Kraft dafür zu arbeiten, dass er im November die Wahl gewinnt, hat sie ihm selbst im Moment seines Triumphes noch Medienaufmerksamkeit abgejagt.

Außerdem holt, wer Hillary Clinton an Bord nimmt, ihren Ehemann Bill mit dazu. Das ist nach dessen Ausfällen während der Vorwahlen nicht nur unangenehm, sondern tendenziell sogar schädlich. Es würde auch Obamas Glaubwürdigkeit schaden, der doch stets von Veränderung und politischer Runderneuerung Washingtons spricht.

Demokratische Stammwählerinnen muss Obama nicht überzeugen. Ob jene aber, die für Hillary waren, weil Obama ihnen ein Graus ist, mit ihr als Vize plötzlich für ihn stimmen würden, ist ungewiss. Umgekehrt dürfte er all jene verlieren, denen die Clintons seit Jahren ein Hassobjekt sind.

Obama muss die Quadratur des Kreises gelingen: Clinton leidlich ruhig stellen, ohne sich erpressen zu lassen, und ihre Fans für sich zu gewinnen. Leicht ist das nicht. Aber ein Vorgeschmack auf kommende Aufgaben.

verboten

Guten Tag,
meine Damen und Herren!

Anstatt endlich den Hanf freizugeben, geht ein aktueller Vorschlag der niedersächsischen Grünen in eine ganz andere, absolut beknackte Richtung. Demnach sollen bei Kommunal- und Landtagswahlen schon 14-Jährige ihre Stimme abgeben können, denn: **„Wir wollen diejenigen, die von heutigen Entscheidungen am langfristigsten betroffen sind, besser beteiligen."**
Nach dieser Logik müsste allen Bürgern spätestens im Rentenalter das Wahlrecht entzogen werden, weil sie die nächste Legislaturperiode eh nicht überleben. Gäbe man allerdings 14-Jährigen politische Verantwortung, unsere Bundeskanzlerin könnte künftig ganz anders heißen:

Prinzessin Lillifee

übrigens: verboten darf nicht *tagesschau* heißen

Figure 5.1 Front page of the German newspaper *die tageszeitung*, June 5, 2008: "Onkel Baracks Hütte" ("Uncle Barack's Cabin").

Figure 5.2 Scan path of a recipient reading the front page of the German newspaper *die tageszeitung*, June 5, 2008: "Onkel Baracks Hütte" (see Figure 5.1). The numbers indicate the sequential order of the fixations.

approach, the concept of "multiplication" as a metaphor for multimodal meaning making is somehow misleading: multiplication implies that the elements that are multiplicated are already identified and fixed. Otherwise, we could not calculate the product. But this condition is not given in multimodal communication: with each step in the reception process, the meaning of the single element is expanded and, therefore, the context for the other elements is changed. Only due to this interpretive dynamic does the repeated fixation of the *same* element make sense.

2. Multimodality inherently needs the notion of *nonlinearity*: multimodal forms of communication are not only ensembles of several modes but also exhibit a nonlinear structure for the recipient, be it a picture, a newspaper page, or a presentation: The recipients have to decide which elements are relevant and in which order they should be perceived. This problem of selectivity (see Bucher and Schumacher 2006) also holds for films: besides their temporal structure, films are also spatially organized as "composition of the shots" (van Leeuwen 2005a: 181) or within "phases" which constitute a sequence (Baldry and Thibault 2005: 47). The fact that different viewers can see different things in the same film (as can the same viewer watching it twice) indicates that the selection of the relevant elements is an integral part of understanding and interpreting films (for more details see Section 4.3). Therefore, one can conclude from reception data that multimodal ensembles have a hypertextual or hypermodal structure.

Speaker- or producer-centered approaches to multimodality seem to imply that each multimodal utterance "has" the meaning that the author intends and that all recipients understand somehow as intended by the utterer. Of course, we know from all different forms of communication—be it a face-to-face conversation or media discourse—that this is rather implausible. The following example of two scan paths of the same homepage demonstrates that the meaning making of the author has to be distinguished from that of the recipient.

2.2 Example 2: Same stimulus—different scan paths

The scan paths of two recipients exploring the landing page of the online edition of the most influential German tabloid *BILD*, published one day after the terror attacks in London in July 2005, indicate that the recipients apply different strategies to allocate their attention, although for both of them the lead story with picture and headline is the entrance point into the page. One strategy (gray

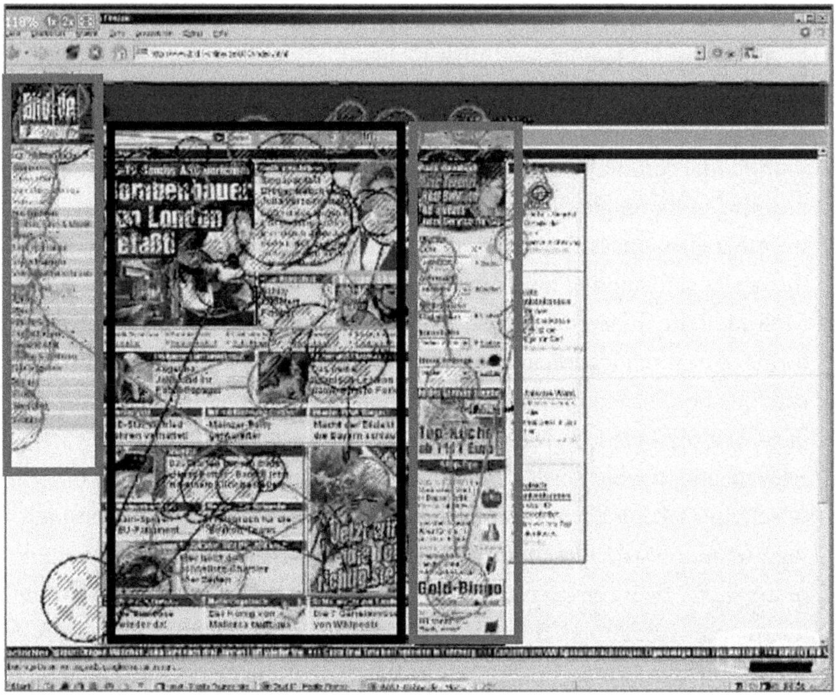

Figure 5.3 Scan paths of two recipients (black and gray) exploring the homepage of the online edition of the German tabloid *BILD*, July 8, 2005.

lines and bubbles) focuses on the navigation tools of the page and the other on its content (black lines and bubbles).

This second example shows that there is no direct unilinear connection between a multimodal stimulus and the sense-making process. This observation is highly relevant with regard to a theory of multimodal meaning making: if participants react differently to the same multimodal document, then recipient meaning cannot be a direct function of the single modal elements but must be a result of their interpretation.

One of the first and most famous eye-tracking studies, which the Russian psychologist Alfred L. Yarbus conducted already in 1967, reached a very similar result for the reception of pictures. Depending on the tasks for observing a painting, different aspects of the picture are relevant, which prompts different patterns of eye movements. Asked to estimate the material circumstances of the depicted persons, the observer's attention is attracted by the environment of the room, while recipients asked to estimate the age of the depicted persons are primarily focusing on their faces. The experiments verified that the intentions with which a picture is observed have a crucial impact on the allocation of attention and, therefore, "the character of the eye movements is either completely independent of or only very slightly dependent on the material of the picture and how it was made" (Yarbus 1967: 190). The author concludes: "Any record of eye movements shows that, per se, the number of details contained in an element of the picture does not determine the degree of attention attracted to this element. This is easily understandable, for in any picture the observer can obtain essential and useful information by glancing at some details, while others tell him nothing new or useful" (Yarbus 1967: 182).

We can easily transfer this idea of selective attention to multimodal stimuli: which modal elements contain "essential and useful information" (see above) is determined not by the orchestration of the multimodal ensemble or the visual salience of an element but by the recipient's intentions. In other words, which elements the recipient acquires is an outcome of the *interaction* between the recipient and the multimodal object.

2.3 Example 3: Prior knowledge and meaning making

Besides their intentions, there is another feature of the recipients that has an impact on the reception process: their *prior knowledge*. Having some schemata or frames at one's disposal facilitates the acquisition of new knowledge. This becomes obvious if we compare how *experts* and *novices* process slides in scientific presentations. The following example comes from an eye-tracking study on scientific presentations (Bucher and Niemann 2012: 300–301). The slide visualizes the results of a medical experiment with rats (see Figure 5.4). The speaker is commenting on the slide with the following words:

> On the left side [Laser pointer on the left part of the picture] you see two pictures of a 'naive' rat [Pointer is following the referring utterances]. Here in this area are the eyes. Here you see the lungs. And this is the visualization of the brain. And one can easily see that in the naive animal Verapamil is, in the end, not absorbed by the brain. (*translated from German*)

This rather complex slide with several areas of interest exhibits high modal density and, therefore, forces the recipients to first select the relevant aspects and then to organize their paths of navigation according to the utterances of the speaker. The eye-tracking data show that the expert and the novice solve these tasks very differently: the expert is able to synchronize his attention with the dynamic of the lecturer's utterances, while the novice is looking—somewhat

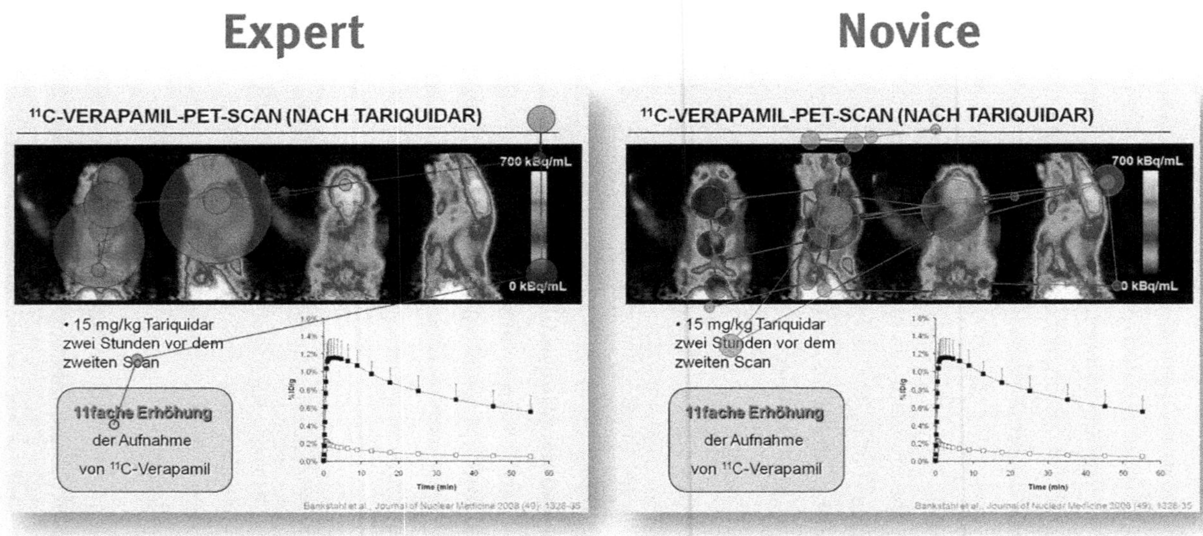

Figure 5.4 Novice versus expert: Two different scan paths of a slide in a scientific presentation.

wildly and arbitrarily—at every part of the picture. Apparently, the expert's higher level of prior knowledge allows him to interpret the verbal references of the presenter—"on the left side," "here," "this"—in the intended way. With the help of his knowledge, the expert is able to focus on the relevant aspects of the slide, while the novice, inexperienced in analyzing positron emission tomography scans, lacks the criteria of relevance.

We can sum up the outcomes of these empirical examples like this: Multimodal meaning making is an iterative and recursive process in which a recipient uses the different modal elements of a multimodal ensemble as contexts for their mutual interpretation. The selection of the relevant elements, the order of their interpretation, and the established relations between them are dependent on the intentions and the knowledge of the recipient. Or, in short: The meaning of a multimodal ensemble is not a function of the modal elements or signs themselves but the outcome of the interaction of a recipient and the multimodal ensemble.

3 Sign-centered approaches to multimodality: The semiotic paradigm

Against the background of the requirements on a theory of multimodality inferred from the reception data, one can compare and discuss some of the central ideas of two approaches to multimodality: the *social semiotic approach* and the grammatically inflected *multimodal discourse analysis*. They are the most influential approaches in multimodality research, and both have great benefits for extending discourse analysis beyond written text and language. The *social semiotic approach* is associated, for example, with Gunther Kress and Theo van Leeuwen, and the grammatically inflected *multimodal discourse analysis* is associated, for example, with Kay O'Halloran, Radan Martinec and Andrew Salway, Anthony Baldry, and Paul Thibault. Both approaches have a similar theoretical background but make different proposals for analyzing multimodal discourse. Both have their roots in the systemic functional grammar of Michael Halliday and refer to semiotics, placing the concept of sign at the center of their theory. Although they use the label "discourse" as part of their name, this concept is much less elaborated than the concept of sign.

Two of Halliday's ideas became particularly strong starting points for the analysis of multimodality: his idea of *metafunctions* and his concept of the *stratification of language* (Halliday and Matthiessen 2004). According to the first approach for an entity to be a sign, it is constitutive, in terms of metafunctions, to have three semiotic resources for meaning making. Every sign expresses experiences and something about the world—its *ideational meaning*—it constitutes a relationship between the sign-user and the addressee—the *interpersonal meaning* —and it contributes to the organization and structure of the text, in which the sign is embedded—the *textual meaning* (Halliday and Matthiessen 2004: 29–31). A basic assumption of both sign-centered approaches to multimodality is that one can apply the categories developed for the analysis of language to all other modes, presupposing that all signs have the same meta-semiotic resources.

The second idea of Halliday that was transferred to multimodal analysis is his concept of *stratification of language* in three layers or dimensions. Every text can be analyzed in the dimension of its context, its content, and its expression (Halliday and Matthiessen 2004: 24–28). An important connector between the different layers is the concept of "genre" or text type: Genres link the text with a special context and regulate the adequate selection of semiotic means dependent on that context (van Leeuwen 2005b).

Despite their merits of placing the modal diversity of communication in the center of research and thereby extending the concept of meaning to all different types of signs, these two approaches exhibit some crucial problems if reviewed in terms of multimodal comprehension or investigated from an action-theoretical theory of discourse. These problems are closely related to their concept of sign and their being anchored in structurally inflected semiotics. Therefore, a discussion of these problems can be taken as an adequate strategy to prepare a consequently more discourse-based and action-theoretical perspective as an alternative to the semiotic one. The two central problems are, first, the *reification of intermodal relations* and, second, the *overemphasis on sign making.*

3.1 Problem 1: Reification of intermodal relations in multimodal discourse analysis

For the analysis of intermodal relations, the different approaches propose lists of different types of relations like "elaboration," "extension," "enhancement," "co-contextualisation," or "re-contextualisation," taken from Halliday's types of relationship between clauses (Halliday and Matthiessen 2004: ch. 7; Martinec and Salway 2005; Liu and O'Halloran 2009; Martinec 2013). Rhetorical Structure Theory, which, for example, John Bateman refers to, distinguishes even more than twenty different relations between modal elements, among them are relations like "Evidence," "Background," "Evaluation," "Summary," or "Justify" (Bateman 2008: 141–51). When describing a relation between two elements, as for example an "extension" in the sense of Halliday as "adding some new element" (Halliday and Matthiessen 2004: 378), this analysis presupposes that one can interpret the two elements in isolation: Only if I have understood element 1 and element 2, am I able to judge that the second has added new information to the first. We can run through this argument for each of the proposed types of relation, and we will end with the same result. From this *reification of modal relations* follows that the system-functional approach to intermodal relations presupposes that the problem, which should be tackled, is already solved. The types of relations are not the *explanans* but the *explanandum*. Catalogues of intermodal relations cannot solve the problem of compositionality in multimodal discourse; they do not even give us criteria to select the elements that are relevant for a coherent interpretation. Besides these *theoretical* problems, the reliance on a separate interpretation of individual modal elements in isolation from their respective contexts is in conflict with the *empirical* results from eye tracking, which confirm a mutual interpretation of modal elements in a multimodal ensemble. The zigzag line of eye movements between different modal elements indicates that multimodal understanding is not a result of adding up the somehow fixed meanings of single elements but the result of a recursive process of interpretation, in which the elements contextualize each other.

Taking into account the logic of the listed relations of intermodal relations, it seems quite plausible *not* to interpret them as grammatical or as propositional relations but as *communicative actions*:

- to *elaborate* a description of A by showing a picture of A;
- to *extend* a report by means of an information graphic;
- to *justify* a comment by citing arguments of an expert;
- to *mark* an article as lead story by placing it at the center of the title page; and
- to *dramatize* a video sequence by adding a piece of music.

These phrases are not descriptions of relations between elements of different sign ensembles, but descriptions of actions performed by a communicator using these elements in combination. In this sense, the deeper layers of intermodal relations are patterns of actions that are performed by using the communicative resources of the modal elements. This suggests that a theory of communicative action is a more fundamental and more natural approach to multimodality and especially to the description of intermodal relations than a codal compositional theory of signs.

The advantage of this approach can also be demonstrated with respect to the analysis of the meaning of single nonverbal modes like, for example, spatial composition in the mode of visual design. For the analysis of newspaper pages, Kress and van Leeuwen suggest a kind of visual grammar of particular spatial configurations, each of which attributes a certain information value to the respective informational element (Kress and van Leeuwen 1998). According to linguistic accounts of metafunctions, a position on the left side of the page expresses that the information is "Given"—already known to the reader—and a position on the right side of the page expresses that the information is "New." Polarizing layout positions vertically, elements placed at the top are presented as the "Ideal," and those placed at the bottom as the "Real" (Kress and van Leeuwen 1998: 189–90) First of all, there is little empirical evidence for this visual grammar and its assumption of connections between compositional zones and their meaning. For example, on the front pages of German newspapers, it is just the other way round. The "New" is mostly positioned on the left side—the table of contents or the brief box—and the "Given" on the right side—the editorial or commentary article, which already presuppose the issue to be commented as well established (see Blum and Bucher 1998). Above all, eye-tracking data indicate that recipients obviously apply quite different "visual grammars" when navigating newspaper pages or web pages (see, e.g., Figure 5.3). Of course, compositional choices in the layout of a page or a picture are one of the meaning-making layers of multimodal discourse. But one should not analyze this modal resource in terms of a fixed and immediate relation between the form and the function or as a codal relation between a signifier and a signified (see also Bateman 2008: 40–53). Since design can be characterized as a parasitic mode, which can only be deployed in combination with other content-based modes, the meaning of page composition as well as the meaning of features of film design like camera angle, shot, or editing can only be inferred in the context of other modal elements. In opposition to a decoding model, an inferential approach to meaning, based on a theory of communicative action, is presented in Section 4.

3.2 Problem 2: Overemphasizing sign *making*

The concept of sign making is central in a social semiotic theory for two reasons. First, it allows one to explain the dynamics of communication or discourse, and second, it is the presupposition for interpreting the signs used as traces of the intentions, the ideological or the cultural background of the communicator. To quote Gunther Kress, "The focus on sign-making rather than sign use is one of several features which distinguish social-semiotic theory from other forms of semiotics" and "signs are made rather than used" (Kress 2010: 54). From this proposition he concludes: "A theory of use is redundant in an approach which has sign-making and the sign-maker at its center: the sign after all is made in and for the conditions of its use" (Kress 2010: 61–62).

The theoretical generalization of this conception, which he contrasts with Saussurean semiotics, is rather radical: "In social semiotics arbitrariness is replaced by motivation, in all instances of sign-making for any kind of sign" (Kress 2010: 67). Kress illustrates his "theory of the motivated sign" with an interpretation of a drawing by a three-year-old child, with the title, "This is a car."

Without knowing the title and having no background information about the creator of the drawing, one most likely will not be able to give an adequate interpretation of this image. A lot of context and a lot of mutual knowledge of the participants of this sign exchange are required to understand this drawing. Obviously it is not as easy as Gunther Kress analyzes situations like this: "In Social Semiotics, if I want to be understood, by preference I use the resources that those around me know and use to make the signs, which I need to make. If I am not familiar with those resources, I make signs in which the form strongly suggests the meaning I want to communicate" (Kress 2010: 64).

Without knowing the context of the drawing, the circle forms in Figure 5.5 suggest very different things: rings, candies, bugs, fish, wheels, coins, etc. The interesting part of Gunther Kress' description is "the resources that those around me know and use to make the signs, which I need to make" (see above). This means without reference to rules, regularities, common practices, or principles for the use of signs, it is not possible to explain mutual understanding by using

Figure 5.5 *This is a car*. Drawing by a three-year-old boy (Kress 2010: 55, Figure 4.1).

signs. The core of truth of Kress' idea lies in the assumption that in a theory of discourse the concept of communication is more fundamental than a system of signs. Tomasello (2008) worked out this approach with respect to the origins of human communication, arguing for the logical priority of communication over language: A joint frame of attention and shared intentionality constitute the psychological infrastructure of human communication, which is essential for all forms of cooperation and communication.

Social semiotics obviously overemphasizes or overgeneralizes the aspect of sign making. Granted, there are cases in which it makes sense to speak of sign making in a literal sense: for example, in communication with the help of gestures; if we do not speak the language of an addressee; or if a university is designing a new logo that should express its corporate identity; or if a child draws circles to visualize a car; and so on. But a lot of signs are already made—like linguistic signs or even design patterns—and we use them to communicate. Of course, language always provides us with alternatives, and one's selection of verbal means in meaning making could be an indication of his or her ideological intentions. But selecting a sign is not the same as "making" it, because selection presupposes that the selected element is already on hand and one is familiar with its meaning potential. In German, for example, there exist two terms for a person applying for asylum: "Asylbewerber" (neutral: "applicant for political asylum") and "Asylant" (negative: "misuser of political asylum"). The question of which of the terms one uses makes a big ideological difference. The same applies to design, color, or typography. Selecting one of the semiotic alternatives on hand for achieving a communication goal presupposes that these alternatives have mutually well-known rules of use we can rely on and that we can assume that the addressees rely on, too. In media communication as well as in everyday talk, it would be stressful and exhausting for communicators and addressees, if there were no routines, rules, principles, or standard cases for the use of signs and instead every sign had to be made situationally.

Arbitrariness and motivation do not exclude each other because they operate on different levels: Arbitrariness refers to the sign, and motivation refers to the use of signs in communication. Therefore, a theory of use is not redundant, as Kress claimed, but a centerpiece of a theory of multimodality. In Figure 5.1, the phrase "Onkel Baracks Hütte" ("Uncle Barack's cabin") or the photo of the White House are, in a way, arbitrary: "Onkel" ("Uncle"), "Barack," or "Hütte" ("cabin") can be used to refer to very different things, but the allusion to the famous novel is a highly motivated use of this phrase, presupposing a very special context of communication. The White House can be represented by very different pictures as well. But the compositional use of all these signs in the multimodal ensemble of a title page is highly motivated, far beyond routine newspaper journalism. As these rather simple examples already demonstrate, analyzing multimodal compositions requires a rather sophisticated theory of use, which goes beyond decoding individual signs.

The problem mentioned by Gunther Kress, that it is the motivation or the intention of a sign-user which is relevant in communication and not the form of signs, can be solved very smartly by referring to an inferential approach, which was prominently proposed by Paul Grice (1975). His concept of *implicatures*, which helps bridging the gap between sentence-meaning and utterance-meaning (or "sense"), can be transferred easily to multimodality to explain cooperation and communication with all types of signs. As a consequence of the problems of sign-centered semiotic approaches to multimodality mentioned so far, the next section presents some approaches that are based on an inferential concept of meaning making and on a pragmatic, action-theoretical approach to discourse and communication.

4 From sign to communication: Pragmatic approaches to multimodality

In contrast to the sign-centered theories of multimodality mentioned before, there is a family of pragmatic theories that assign the notions of communication and discourse a central role. Compared to sign centered approaches, these discourse-based approaches imply "a more indirect relationship between material traces and attributions of meaning" (Bateman, 2016: 44). They share the assumption that communication should not be analyzed as an exchange of signs but as a form of cooperation for which the different types of signs are used as means to an end. Accordingly the term "pragmatic(s)" is used here in a fundamental sense: not to denote an additional core component of linguistic theory besides phonetics, syntax, and semantics but as a superior perspective in which language and other sign systems are seen in their use in communication as a social practice. In this sense pragmatics is "logically prior to semantics" (Levinson 1983: 35), as we have access to signs and their meaning potentials only via the use of signs in communication (see also Gee and Handford 2012; Huang 2014: ch. 1). With this perspective on multimodality, the central unit of analysis is neither the sign maker nor the sign itself but the interaction between communicator and addressees, for which the modal resources are used. Intermodal relations between text and image, between sound and film, or between design and article are analyzed not as constellations of signs but as complex discursive units addressed with a special intention to an audience or a person. In line with this conception, the common ground of different modes is not to be located in semantic metafunctions but on a higher level of communication, that is, in discourse patterns like narration, explanation, argumentation, or advertising. Pragmatic, discourse-based theories also share the assumption that comprehension and interpretation of multimodal discourse cannot be explained as a matter of *decoding* a somehow fixed relation between sign and meaning but as an *inference* from the sign in use and its discernible features on the basis of mutual knowledge and its rule-based or routine usage. The more abstract level for analyzing multimodal relations can either be a theory of discourse (Bateman 2014; Bateman and Schmidt 2012; Bateman and Wildfeuer 2014), a theory of cooperation (Forceville 2014), or a theory of communicative actions (Bucher 2007, 2011a, 2012; Fritz 2013).

The following subsection will demonstrate how the two basic problems of multimodality research mentioned before—the problem of compositionality and the problem of reception—can be solved by applying a discourse-based theory without falling into the pitfall of reifying intermodal relations or overemphasizing meaning making.

4.1 Multimodal discourse as a form of cooperation

From the perspective of cooperation theory, Forceville (2014) defines multimodal communication as "ostensive-inferential": "It is clear to both sender and addressee that the sender wants the addressee to be aware that she directs a message to him and that he is to infer relevant information from this message" (Forceville 2014: 53). This "shared intentionality" (Tomasello and Carpenter 2007) is the constitutive condition for the possibility of communication in general and, therefore, the precondition for the process of inference being triggered by the signs used to perform the utterance. As mentioned before, the idea that the meaning of an utterance is implicit and inferential and not an explicit feature of the signs used in communication goes back to the philosopher Paul Grice, whose conception explains "a link between intention and meaning and consequently of a logical connection between 'cooperation' and the transmission of information"

(Black 1975: 136). Essential in his approach is the concept of "conversational implicature," which refers to the implicit meaning of an utterance (Grice 1989). To explain the inferential process in communication, Grice proposes a cooperative principle and four maxims that govern all communication (Grice 1989: 26–27): the *maxim of quantity* (make your contribution as informative as is required); the *maxim of quality* (do not say what you believe to be false or for which you lack evidence); the *maxim of relation* (be relevant); and the *maxim of manner* (be perspicuous, avoid obscurity, ambiguity, and be brief and orderly). These maxims are not conceived as normative demands but as constitutive conditions for the possibility of communication generally to serve "a maximally effective exchange of information" (Grice 1989: 28). Hence, it is consistent to label these maxims as "conversational imperatives" (Grice 1989: 370) or as "the psychological infrastructure of human communication" (Tomasello 2008: 107).

Against the background of this infrastructure, we can explain multimodal understanding without getting into the difficulties of mixing up arbitrariness and motivation: arbitrary signs of different modes can be used to bring the addressee to understand the intentions and motivations of the speaker by applying the cooperative principle and the conversational maxims. The addressee presupposes that the speaker is cooperative and has combined the modal elements in a meaningful and comprehensible way, that each of the modal elements is relevant, that their orchestration is coherent and perspicuous, and that the speaker has enough evidence for the information expressed. The inferential process detecting the conversational implicatures of a multimodal discourse is guided by these maxims and can be triggered by salient features of the modal elements, either textual or material.

Figure 5.1 is a good example, demonstrating the suitability of this inferential approach to multimodality: It is obvious that the meaning of the lead story in Figure 5.1 cannot be deduced by adding the meaning of the single elements because neither the picture nor the headline nor the lead text unfolds its full meaning in isolation but only in the context of each other. Therefore, the meaning of this multimodal orchestration is not explicit and somehow deducible directly from some salient features of the single textual or pictorial elements, but implicit and only accessible inferentially. Apart from the mutual knowledge concerning the novel *Uncle Tom's Cabin* and the photograph depicting the White House, recipients have to presuppose that the newspaper journalists are following the cooperative principle and the conversational maxims, although the headline "Uncle Barack's Cabin" sounds somehow irrelevant and uninformative in the context of the coverage of the US presidential primaries. Due to the assumption that the journalist is following the cooperative principle and the conversational maxims, an inferential process is triggered by this anomalous utterance. The recipients will assume that the journalists intend to communicate with their audience, that the multimodal orchestration is not obscure but perspicuous, that the article is about something that really happened and of which the journalists have evidence (namely that Obama was elected as candidate of the Democratic Party), that each of the modal elements is relevant, and that the content of the article is informative for the newspaper reader under the given temporal circumstances. Together with the relevant prior knowledge, these maxims can lead the newspaper reader to the conclusion that the article reports about Obama having won the Democratic primary and at the same time accentuates the historic dimension of this election by alluding, with a play on words, to one of the most famous antislavery novels. The photograph of the White House links the historical dimension with the present age and visually triggers another allusion to black-and-white symbolism. Far beyond this example, it is the concept of conversational implicature that can explain how and why comprehension of multimodal discourse is possible, even if the meanings of iconic elements like visuals, gestures, design, or sound are not defined in a dictionary but must be worked out from the discourse itself.

4.2 Multimodal discourse as complex communicative action

Empirical research on multimodal discourse from very different approaches often refers to categories like "describe," "elaborate," "justify," "advertise," "narrate," etc. in order to characterize the function of modal elements. These references to categories of action suggest that a theory of communicative action is quite a natural approach to multimodality and especially to the description of intermodal relations. Applying this approach systematically offers some helpful tools for analyzing the complex semiotic architecture of multimodal discourse. Therefore, one can refer to three different types of combination of communicative actions (Goldman 1970; Heringer 1978: ch. 2):

1. Actions can be linked as a sequence in an "*and-then-relationship*":
   A newscast reports an event *and then* shows on a map where it happened.
2. Actions can be linked as a constellation in an "*and-simultaneously-relation*":
   In a newspaper article, a political decision is reported *and simultaneously* the design indicates that the article is the lead story of the day.

3. Actions can be combined in a hierarchy (the "*by-relation*"): One can perform a complex higher-level action "*by*" performing several less complex lower-level actions. For example, a newspaper journalist can portray a politician *by* describing his political career, *by* showing a picture of his family, and *by* presenting his personal data sheet in an information box (see Goldman 1970: 20).

Taking these three types of relationship together, we have a tool that allows us to describe the temporal and spatial structures of even complex multimodal ensembles. Table 5.1 demonstrates how one can analyze a newspaper article that is the lead story of the day and that is presented as a multimodal ensemble combining visuals, design, layout position, and text:

The left column contains the action at the highest level, expressing the fundamental function of the article to inform an audience; the third column contains the subordinate actions of a second and third level, which are necessary to achieve the basic intention of the article; column two lists the types of relation between the actions on the first and the following level, whereas column four contains the modal elements used to perform this complex multimodal action. Of course, this is a simplified description of the structure of a complex multimodal action in several respects. First, the string of actions can be extended to the left, as the action to inform about X could be part of a superordinate

Table 5.1 Structure of a multimodal action (A = newspaper, B = audience, X = issue of the article)

A informs B about X	By	reporting how X happened	Text 1
	By	showing how/where X happened	Photo, Graphic, Video (on a website)
	and simultaneously	indicating the news value of the information by placing the article on the title page	Design
	and simultaneously	indicating which elements belong to the report about X	Design
	and then	commenting the reported event	Text 2

action, for example "supporting a political party" or "continuing an issue of the day before." Second, column three can contain more detailed subordinate actions than listed; and third, the string of actions can be extended to the right side to make the analysis even more adequate by attaining a higher level of granularity. For example, A can show where X has happened *by* using a map *and simultaneously* indicating exactly where X happened *by* positioning a red flag in the map. In sum, the table elucidates the somehow obscure concepts of "fusion," "marriage," or "multiplication," which are used in sign-centered approaches to multimodality to describe intermodal relations. In an action-based approach, intermodal relations are relations between actions on different levels in a hierarchy, which represent the complex structure of a multimodal orchestration.

4.3 Analyzing intermodal relations

Considered together, a theory of cooperation and inferential meaning and a theory of communicative action can be used as complementary building blocks of a theory of multimodality. The following example from a reception study demonstrates how these two approaches collaborate in solving the central problems of any theory of multimodality mentioned above, namely the problem of compositionality and its mirror image, the problem of comprehending multimodal relations. The example comes from a study that investigated the reception of a commercial with about sixty test subjects (Bucher 2011b, 2012). The commercial promotes a new mobile phone by telling the story of a person watching a soccer game on this mobile phone, which makes him so absentminded that he forgets that he is sitting in the lounge of a hotel and not in his living room at home. To identify the influence of knowledge and intentions, the study was conducted with different recipient groups who got different tasks for watching the video clip—in analogy to the experimental setting Yarbus (1967) applied (see Section 2). To investigate the problem of compositionality and to identify the influence of single modes on the interpretation of the video, some of the groups saw the videos *with* sound and some *without*. This experimental manipulation of switching off a modal layer was used to allow for isolating the meaning potential of the soundtrack of the film—both sounds and spoken language.

How important the soundtrack is for understanding the commercial becomes obvious if we compare the reports of the two groups. The subjects who saw the film with sound give a *re-narration of the story* while the "no sound" group only *describe what they saw*. The "sound" group is able to tell a story from the perspective of the protagonist ("The protagonist is so fascinated by the TV program . . .," "he forgets that he is not at home"), their stories contain relational sequences, which explains the coherence between single episodes, and they draw a conclusion on the full story in formulating the overall intention, the point of the spot ("This means: mobile TV gives you the same TV experience as watching at home"). Subjects from the "no sound" group describe in particular what they saw ("One sees a good looking man"; "In a close-up his face is presented"; "then a zoom shows the scenery"), aspects that are hardly mentioned by subjects from the "sound" group. In addition, the reproductions of the "no sound" group display a distinct additive structure (and then . . . and then), restricted to temporal relations between the episodes ("And then he presumably got a phone call and then he realizes that he is eating flower soil"). One of the test persons of the "no sound" group formulates problems that were caused by the absence of the aural modes, particularly that she could not hear the phone ringing in the key episode of the video: "Until he talks with the waiter I was somehow able to understand the plot. But from this point I did not get why he picks up his mobile and why the video ends this way" (all recipient quotations here are translated from German).

This comment indicates that the problem of understanding and orientation is brought about by the missing clues from the aural mode of sound. And indeed, the relevance of the modal resources of the soundtrack for directing the recipients' attention becomes obvious when we compare the eye-tracking data of the groups with and without sound.

The upper part of Figure 5.6 documents the eye-tracking data of the group *with* sound; the lower part the data of the persons watching the commercial *without* sound. The different colors mark different areas of interest for which the fixations were measured. Fixations are the time span during which the eye remains still and processes the visual information of a specified area of interest. For example, the dominating black parts represent the protagonist of the spot and the grey parts between the two vertical lines represent the waiter as an area of interest. Comparing the pattern of the eye movements of the two groups reveals some peculiarities of the reception without sound:

- The fixation on the waiter starts later (on average 160 ms).
- The absolute time of fixation on the waiter lasts longer (on average 216 ms).
- The shift of attention from waiter to protagonist happens later (on average 160 ms).
- There are about twice as many fixations on the cell phone.

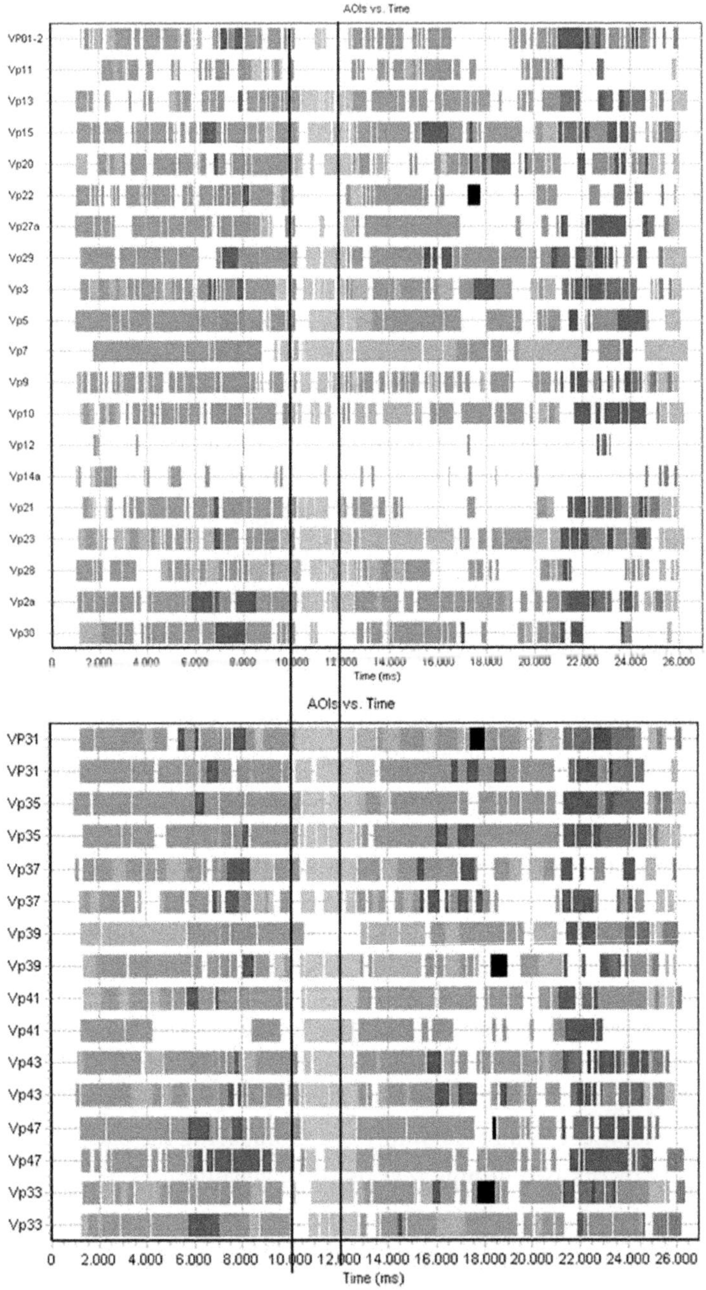

Figure 5.6 Eye-tracking data of twenty test subjects of the "sound" group (above) and twenty test subjects of the "no sound" group (below). The bars with the different colors indicate the areas of interest and the duration of its reception. The light gray bars between the two vertical lines indicate the fixation on the waiter.

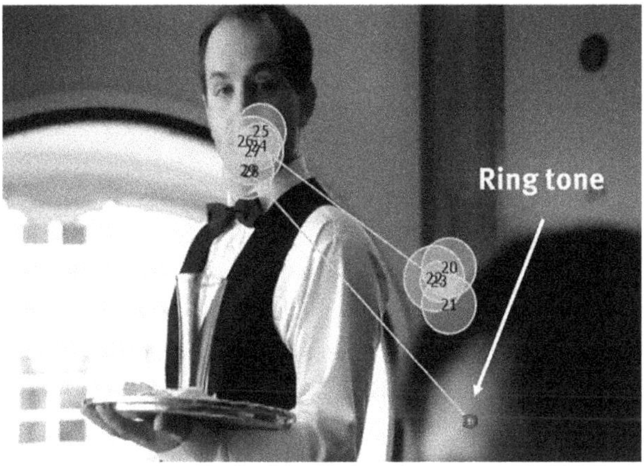

Figure 5.7 Influence of sound—the ringing of the phone—on recipients' eye movements.

The differences between the eye-tracking data of the two groups document that the aural mode influences the allocation of attention and the perceptual processing of the visual field. These differences allow us to reveal the modal resources of sound and its function in audiovisual media. Figure 5.7 verifies this function by demonstrating the influence of the ringing of the phone on the eye movements of a recipient:

The moment the cell phone rings, the attention shifts away from the waiter. This is an important observation because a salience-based bottom-up theory focusing on visual elements cannot explain why the eyes of the protagonists are moving to an area of the screen where nothing significant can be seen. This area is only relevant in the context of the story told: the recipients assume that it is the protagonist's cell phone ringing. In the case of the subjects who see the silent movie, the attention does not shift away from the waiter until the protagonist is completely on screen.

This short episode in the film is of high significance for getting the point of the story across. The ringing of the mobile phone makes the protagonist realize that he is not watching the football game at home but in a restaurant. This switch from the subjective reality of the protagonist to the objective reality of the narration is staged in the film by a rather complex intermodal arrangement of visual elements and sound. None of these co-deployed modes is read in isolation, but each is mutually interpreted with the help of the others. Neither the ringing of the mobile phone alone nor the simultaneously presented video sequence makes sense on their own, but together they constitute an episode within the story told in the commercial. The ringing indicates what element in the visual mode of the video is relevant in this episode, namely the mobile phone and not the waiter. The sound directs the selection of the elements that are relevant in the visual mode—although they are not even seen. Furthermore, the ringing of the phone is an important clue for the interpretation of the protagonist's facial expression.

Understanding the ringing of the phone, therefore, does not mean simply decoding an aural sign, although the spot presupposes that the recipient knows in principle what the ringing of a phone means, namely that it indicates that someone gets a call from another person. But this knowledge is only the starting point for the intended deeper interpretation. The ringing and the facial expression of the protagonist constitute the episode of the story told in the commercial in which the protagonist realizes the difference between his subjective reality and the objective reality of the scene. To understand the interrelations of the different modal elements, one has to integrate the actions performed with the different modes—visual and aural—into a superordinate action of communication: telling the story of a person who watches a football game on his mobile phone in a restaurant, assuming that he is at his home. And, of course, understanding the video presupposes that we integrate this narrative discourse into another higher-level structure of advertising a new cell phone. As demonstrated in the case of the newspaper article in Figure 5.1, the inferential meaning making in this example as well refers to the Gricean cooperation principle and conversational maxims in several respects: only the assumption that the ringing of the phone is relevant for the story being told prompts the recipients to infer its conversational meaning as marking the reality shift of the protagonist, which is far beyond its conventional meaning of indicating that the protagonist gets a call from another person. Figure 5.8 visualizes the analysis of the multimodal action in which the ring tone is integrated.

Although this visualization simplifies the structure of the video sequence in several respects, it can help explain *the intersemiotic hypothesis* that the meaning of a multimodal ensemble is more than the sum of the meanings of its modal elements. "More" in this case means that the elements of a lower-level action are integrated in a higher-level action. In Figure 5.8, the respective additional meaning ("more") of the aural mode (ringing) and the visual mode (facial expression) is to be found in the boxes left of the last one. What other theories of multimodality metaphorically call "marriage," "fusion," or "multiplication" of modes can be described in an action-theoretical approach as the embedding of

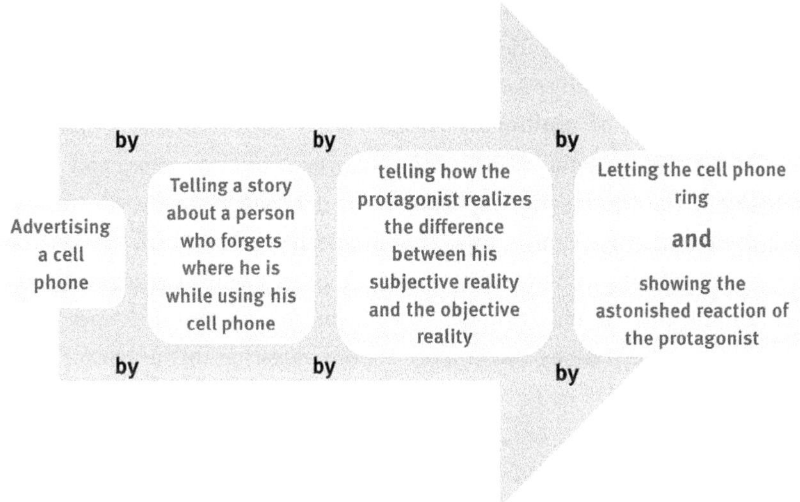

Figure 5.8 Structure of a multimodal action in a commercial for cell phones.

modal elements—like the ringing of a phone or a facial expression—in a higher-level and more complex action. The typical relations for combining actions—the *by-*, the *and-then-*, and the *and-simultaneously-*relations—are the tools for describing the structure of those complex actions. Having more modes at one's disposal means not only that we can communicate via more channels but also that we can perform more actions, more *complex* actions, and *other* actions. The story told in the commercial could not be told in this way with written text or verbal narration because the simultaneous ringing of the cell phone and the facial reaction of the protagonist cannot be expressed in a linear form of discourse but presupposes the co-deployment of visual and aural modes.

5 Conclusions

Analyzing meaning making from a recipient's perspective in the light of empirical data has revealed how one can solve the two basic problems of any theory of multimodality—the *problem of compositionality* and its counterpart, the *problem of reception*. A precondition for this solution lies in replacing a sign-centered approach with a discourse-based and pragmatic approach to multimodality, which helps us avoid the two pitfalls of *reifying intermodal relation* and *overemphasizing sign making*. Additionally, a discourse-based approach makes available theoretical traditions that provide the conceptual means to replace an unsatisfactory decoding model of meaning making with an inferential model. A theory of communicative cooperation provides the opportunity for explaining implicit meanings of multimodal discourse as conversational implicatures with the help of cooperative principles, conversational maxims, and mutual knowledge. The systematic explication of the relation between meaning resources and discursive meaning in this approach prevents one from confusing arbitrariness of signs with intention or motivation of modal utterances. A theory of communicative actions provides the means to describe the structure of multimodal orchestrations as complex actions referring to the basic types of relations between actions like the "*and-then-*," the "*and-simultaneously-*," and the "*by-relation*." These relations can replace the metaphorical and somewhat obscure concepts of "fusion," "marriage," or "multiplication" that are used in sign-oriented approaches to multimodality to describe intermodal relations. In an action-based approach, intermodal relations are treated as relations between actions on different hierarchical levels, which represent the complex structure of a multimodal orchestration.

A discourse-based or pragmatic perspective has some further consequences for a theory of multimodality. The semiotic resources of signs can be analyzed as the inventory of communicative actions one can perform by means of these signs. It is obvious that this functional approach to meaning works for all types of signs, sound, pictures, design, music, and color, as well as for linguistic units. In analogy to language, signs from all modes can be used to pursue communicative aims. In analogy to the concept of sign making in social semiotic approaches, it is assumed that a distinctive feature of all signs is their ability to be used in a context of communicative actions to make the addressee understand something (Keller 1995; Forceville 2014). A set of metafunctions of all signs does not need to be specified but consists of their potential to be used for performing social actions. The fundamental distinction between the meaning potential of a sign, which determines its action possibilities ("utterance type meaning"), and the communicative meaning created by an actually performed multimodal action ("utterance token meaning") prevents this approach from producing a short circuit between sign and communicative meaning.

Furthermore, an inferential approach provides a satisfying explanation of what it means to understand multimodal discourse. An inferential model is based on the selection of relevant elements from the co-deployed modes and reflects the characteristic features of this type of discourse: its nonlinearity and the co-deployment of different modes. Therefore, multimodal understanding has some similarities with navigating a hypertext: each recipient has to find his path of reading, be it a newspaper page, a website, a photo, a film, or an exhibition in a museum. To find a path means, first, to select and classify the elements which are relevant and, second, to assign coherence to the selected elements, which means bringing them under a higher-level pattern of action. Hence, understanding multimodal discourse happens step by step and is not a type of decoding. A modal element attracting attention—for example the headline "Uncle Barack's Cabin" or the ring tone of a phone—actuates a problem of understanding, which the recipient tries to solve with the help of a second modal element—the enclosed photo of the White House or the facial expression of the protagonist (see Figure 5.1). Combining the meaning of the two elements under a coherent interpretation, one achieves a deeper degree of understanding—the next level of interpretation. This process is repeated until the recipient arrives at a sufficient degree of comprehension. As this is an active and recursive process, it can be modeled as an interaction between the recipient and the multimodal arrangement, whose structure is very similar to that of a conversation or dialog.

An aphorism from Greek philosophy attributed to Heraclitus asserts that one cannot step into the same river twice. The same is true for perceiving multimodal arrangements like reading a newspaper, navigating a website, or watching a commercial. With each gaze our knowledge, attention, and attitude change and the next gaze is based on new presuppositions and conditions. This process of interpretation has the structure of an interaction between the recipient and the medium. The recipients act as if the medium is a partner who continuously provides the necessary information for deeper comprehension.

Acknowledgment

The author would like to thank Gerd Fritz for useful suggestions and corrections to earlier versions of this article.

References

Alston, W. (1964), *Philosophy of Language*, Englewood Cliffs, NJ: Prentice Hall.

Austin, J. L. (1956/1957), "A Plea for Excuses: The Presidential Address," *Proceedings of the Aristotelian Society, New Series*, 57: 1–30.

Austin, J. L. (1962), *How to Do Things With Words*, Oxford: Oxford University Press.

Baldry, A. and P. J. Thibault (2005), *Multimodal Transcription and Text Analysis: A Multimedia Toolkit and Coursebook*, London/Oakville: Equinox.
Bateman, A. (2011), "The Decomposability of Semiotic Modes," in K. L. O'Halloran and B. A. Smith (eds.), *Multimodal Studies: Exploring Issues and Domains*, 17–38, New York/London: Routledge.
Bateman, J. A. (2008), *Multimodality and Genre: A Foundation for the Systematic Analysis of Multimodal Documents*, London: Palgrave Macmillan.
Bateman, J. A. (2014), *Text and Image: A Critical Introduction to the Visual/Verbal Divide*, London/New York: Routledge.
Bateman, J. A. (2016), "Methodological and Theoretical Issues for the Empirical Investigation of Multimodality," in N. M. Klug and H. Stöckl (eds.), *Handbuch Sprache im multimodalen Kontext/Handbook on Language in Multimodal Contexts*, 36–74, Berlin: de Gruyter Mouton.
Bateman, J. A. and K.-H. Schmidt (2012), *Multimodal Film Analysis: How Films Mean*, New York: Routledge.
Bateman, J. A. and J. Wildfeuer (2014), "A Multimodal Discourse Theory of Visual Narrative," *Journal of Pragmatics*, 74: 180–208.
Black, M. (1975), "Meaning and Intention," in M. Black (ed.), *Caveats and Critiques: Philosophical Essays in Language, Logic and Art*, 85–108, Ithaca/London: Cornell University Press.
Blum, J. and H.-J. Bucher (1998), *Die Zeitung: Ein Multimedium: Textdesign—Ein Gestaltungskonzept für Text, Bild und Grafik* (Vol. 1), Konstanz: UVK Verlag.
Boeriis, M. and J. Holsanova (2012), "Tracking Visual Segmentation: Connecting Semiotic and Cognitive Perspectives," *Visual Communication*, 11 (3): 259–81.
Bucher, H.-J. (2007), "Textdesign und Multimodalität. Zur Semantik und Pragmatik medialer Gestaltungsformen," in K. S. Roth and J. Spitzmüller (eds.), *Textdesign und Textwirkung in der massenmedialen Kommunikation*, 49–76, Konstanz: UVK Verlag.
Bucher, H.-J. (2011a), "Multimodales Verstehen oder Rezeption als Interaktion: Theoretische und empirische Grundlagen einer systematischen Analyse der Multimodalität," in H. Dieckmannshenke, M. Klemm, and H. Stöckl (eds.), *Bildlinguistik: Theorien—Methoden—Fallbeispiele*, 123–56, Berlin: Erich Schmidt Verlag.
Bucher, H.-J. (2011b), "'Man sieht, was man hört' oder: Multimodales Verstehen als interaktionale Aneignung. Eine Blickaufzeichnungsstudie zur audiovisuellen Rezeption," in J. Schneider and H. Stöckl (eds.), *Medientheorien und Multimodalität. Ein TV-Werbespot—Sieben methodische Beschreibungsansätze*, 109–50, Köln: Herbert von Halem Verlag.
Bucher, H.-J. (2012), "Intermodale Effekte in der audio-visuellen Kommunikation. Blickaufzeichnungsstudie zur Rezeption von zwei Werbespots," in H.-J. Bucher and P. Schumacher (eds.), *Interaktionale Rezeptionsforschung. Theorie und Methode der Blickaufzeichnung in der Medienforschung*, 257–96, Wiesbaden: Springer Verlag.

Bucher, H.-J. and P. Schumacher (2006), "The Relevance of Attention for Selecting News Content. An Eye-Tracking Study on Attention Patterns in the Reception of Print- and Online Media," *Communications. The European Journal of Communications Research*, 31 (3): 347–68.

Bucher, H.-J. and P. Schumacher (eds.) (2012), *Interaktionale Rezeptionsforschung: Theorie und Methode der Blickaufzeichnung in der Medienforschung*, Wiesbaden: Springer Verlag.

Bucher, H.-J. and P. Niemann (2012), "Visualizing Science: The Reception of Powerpoint Presentations," *Visual Communication*, 11 (3): 283–306.

Duchowski, A. T. (2007), *Eye Tracking Methodology: Theory and Practice*, 2nd edition, London: Springer.

Fei, V. L. (2004), "Developing an Integrative Multi-Semiotic Model," in K. L. O'Halloran (ed.), *Multimodal Discourse Analysis: Systemic Functional Perspectives*, 220–46, London/New York: Continuum.

Forceville, C. (2014), "Relevance Theory as Model for Analysing Visual and Multimodal Communication," in D. Machin (ed.), *Visual Communication*, 51–70, Berlin/Boston: de Gruyter.

Fritz, G. (2005), "On Answering Accusations in Controversies," *Studies in Communication Sciences. Special Issue: Argumentation in Dialogic Interaction*, 151–62.

Fritz, G. (2013), *Dynamische Texttheorie*, Gießen: Gießener Elektronische Bibliothek, August 30. Available online: http://geb.uni-giessen.de/geb/volltexte/2013/9243/pdf/FritzGerd_2013.pdf.

Gee, P. and M. Handford (2012), "Introduction," in J. P. Gee and M. Handford (eds.), *The Routledge Handbook of Discourse Analysis*, 1–6, London/New York: Routledge.

Goldman, A. I. (1970), *A Theory of Human Action*, Enlgewood Cliffs, NJ: Prentice Hall.

Grice, H. P. (1975), "Logic and Conversation," in P. M. Cole and J. L. Morgan (eds.), *Syntax and Semantics, Volume 3: Speech Acts*, 41–58, New York: Academic Press.

Grice, H. P. (1968), "Utter's Meaning, Sentence-Meaning and Word-Meaning," *Foundations of Language*, 4: 225–42.

Grice, P. (1989), *Studies in the Way of Words*, Cambridge, MA: Harvard University Press.

Halliday, M. A. K. and C. M. I. M. Matthiessen (2004), *An Introduction to Functional Grammar*, 3rd edition, London: Hodder Arnold.

Heringer, H. J. (1978), *Practical Semantics: A Study in the Rules of Speech and Action*, The Hague/Paris/New York: Mouton.

Holmqvist, K., M. Nyström, R. Andersson, R. Dewhurst, H. Jarodzka, and J. Van de Weijer (2011), *Eye Tracking: A Comprehensive Guide to Methods and Measures*, Oxford: Oxford University Press.

Holsanova, J. (2008), *Discourse, Vision, and Cognition*, Amsterdam/Philadelphia: John Benjamins.

Holsanova, J. (2014), "In the Eye of the Beholder: Visual Communication from a Recipient Perspective," in D. Machin (ed.), *Visual Communication*, 331–55, Berlin/Boston: de Gruyter.

Holsanova, J., H. Rahm and K. Holmqvist (2006), "Entry Points and Reading Paths on Newspaper Spreads: Comparing a Semiotic Analysis with Eye-Tracking Measurements," *Visual Communication*, 5 (1): 65–93.

Holsanova, J., K. Holmqvist, and N. Holmberg (2008), "Reading Information Graphics: The Role of Spatial Contiguity and Dual Attentional Guidance," *Applied Cognitive Psychology*, DOI:10.1002/acp.1525.

Huang, Y. (2014), *Pragmatics*, 2nd edition, Oxford: Oxford University Press.

Keller, R. (1995), *Zeichentheorie: Zu einer Theorie semiotischen Wissens*, Tübingen/Basel: Francke Verlag.

Kress, G. (2010), *Multimodality: A Social Semiotic Approach to Contemporary Communication*, London/New York: Routledge.

Kress, G. and T. van Leeuwen (1998), "Front Pages. The Critical Analysis of Newspaper Layout," in A. Bell and P. Garrett (eds.), *Approaches to Media discourse*, 186–219, Oxford: Blackwell.

Lemke, J. L. (1998), "Multiplying Meaning: Visual and Verbal semiotics in Scientific Text," in J. R. Martin and R. Veel (eds.), *Reading Science: Critical and Functional Perspectives on Discourses of Science*, 87–113, London: Routledge.

Levinson, S. C. (1983). *Pragmatics*, Cambridge: Cambridge University Press.

Liu, Y. and K. L. O'Halloran (2009), "Intersemiotic Texture: Analyzing Cohesive Devices between Language and Images," *Social Semiotics*, 19 (4): 367–88.

Martinec, R. (2013), "Nascent and Mature Uses of a Semiotic System: The Case of Image–Text Relations," *Visual Communication*, 12 (2): 147–72.

Martinec, R. and A. Salway (2005), "A System for Image-Text Relations in New (and Old) Media," *Visual Communication*, 4 (3): 337–71.

Ogden, C. K. and Richards, I. A. (1923), *The meaning of Meaning: A Study of the Influence of Language upon Thought and of the Science of Symbolism*, New York: Harcourt, Brace & World.

O'Halloran, K. L. (2008), "Systemic Functional-Multimodal Discourse Analysis (SF-MDA): Constructing Ideational Meaning using Language and Visual Imagery," *Visual Communication*, 7 (4): 443–75.

Salvucci, D. D. (1999), *Mapping Eye Movements to Cognitive Processes*, Pittsburg: Carengie Mellon University.

Schiffer, S. R. (1972), *Meaning*, Oxford: Oxford University Press.

Tomasello, M. and M. Carpenter (2007), "Shared Intentionality," *Developmental Science*, 10 (1): 121–25.

Tomasello, M. (2008), *Origins of Human Communication*, London: MIT Press.

van Leeuwen, T. (2005a), *Introducing Social Semiotics*, London/New York: Routledge.

van Leeuwen, T. (2005b), "Multimodality, Genre and Design," in S. Norris and R. H. Jones (eds.), *Discourse in Action: Introducing Mediated Discourse Analysis*, 73–93, London: Routledge.

Yarbus, A. L. (1967), *Eye Movements and Vision*, New York: Plenum Press.

Wittgenstein, L. (1977), *Philosophische Untersuchungen (Philosophical Investigations)*, Frankfurt a.M.: Suhrkamp.

Wittgenstein, L. (1980), *Das Blaue Buch (The Blue Book). Eine philosophische Betrachtung (The Brown Book)*, Frankfurt a.M.: Suhrkamp.

6

Approaching Multimodality from the Functional-pragmatic Perspective

Arne Krause

1 Introduction[1]

Multimodality nowadays can be understood in different ways. The most prominent understanding is closely tied to semiotic approaches that vary in their main focus (see different approaches outlined in Norris and Maier 2014). These approaches to multimodality do not focus primarily on language but on various semiotic resources where language is only one among others. In another, primarily linguistic, understanding, the term "multimodality" is borrowed from these approaches to describe a field of research that departs from the idea that language is the only object of linguistic research. This chapter argues from an action-centered linguistic point of view, to show how multimodal research can benefit from this perspective and vice versa.

A question that arises within a primarily linguistic perspective on multimodality is whether multimodality can be approached from a pragmatic perspective. These considerations can lead to research of the investigation of the reception of multimodal data via eye tracking, as Bucher (2011, e.g.; see also Bucher this volume) points out. Another approach is to incorporate the text-linguistic perspective (for an overview of pragmatic approaches to multimodality see Bateman 2014: 223–38), focusing on different objects of research. Thus, it is clear that different theoretical backgrounds can lead to different analytical approaches and findings. This chapter seeks to point out the benefits of one specific pragmatic theory, namely Functional Pragmatics (FP).[2]

FP is a linguistic theory and not, as the compound might suggest, a modification of Morris' (1938) understanding of pragmatics as the usage of signs. Morris' understanding can, following Ehlich (1999), be understood as "additive" pragmatics that views pragmatics as a subdomain of linguistics alongside syntax, phonology, morphology, etc. FP, on the contrary, can be understood as a fully developed linguistic theory that puts the purpose of linguistic actions as historical-societal (cf. Redder 2008) in the center. As pragmatics can be understood as "the *use that speakers and hearers make of signs in context*" (Bateman 2014: 223; original emphasis), FP turns this around: Signs are understood as the resources to fulfill communicative needs, as means of communication. Thus, FP aims to reconstruct the actional quality of means of communication. This is of great importance for the understanding of multimodality that is approached in this chapter, as will be elaborated further.

To elaborate how a functional-pragmatic approach can expand research on multimodality, this chapter is structured into four main parts: First, different perspectives on multimodality will be described and put into perspective. Following this overview, FP will be introduced as a language theory. This will be followed by an outline of methodological tools for transcription, that is, the HIAT system, which is integral for functional-pragmatic analyses. Following the methodological part, a study currently undertaken by the author will be outlined briefly to demonstrate the possibilities of a functional-pragmatic analysis that takes multimodal aspects into consideration.

2 Perspectives on multimodality

Multimodality is a topic of many recent studies from many different perspectives with heterogeneous objects of research. Aside from compelling takes across different areas, like medieval studies (Maxwell 2015), architecture (Siefkes and Arielli 2015), or film studies (Bateman and Schmidt 2012), there is also a plurality of theoretical backgrounds and a discussion whether or not to adopt analytical frameworks developed for non-multimodal data. Machin (2014) points out that analyses of visual semiotic resources should not be conducted with linguistic methodology and theory but with new methodology and theory, developed for this specific purpose—or at least by methodology and theory for visual analysis. While this position is acceptable for visual-only data, it is not when it comes to multimodal data that contain visual *and* verbal elements. Even more so, it is obvious that linguistic data are necessarily multimodal. Hence, an appreciation of linguistic approaches that contains at least a few visual aspects of communication would be wise.

As the recent overview by Wildfeuer (2015) has shown, there are several international discourses on multimodality, some of which are not connected to each other—unfortunately. Moreover, in some cases, an interdependent reception is a desideratum. To narrow one aspect of the discussion down for this chapter: The term "multimodality" itself is in dispute—"multimodality" has become an umbrella term for vastly heterogeneous research. Notwithstanding the roots of multimodal research in systemic functional linguistics (SFL), the core terms of multimodality, that is, "modality" and "mode," have moved far past their original Hallidayan definitions. Modality was at first understood in the traditional grammar sense, that is the facticity of an utterance, mostly realized through modal verbs, the modes being indicative or conjunctive or modal particles. Mode is also central in SFL and linked to an understanding of language as social semiotic and hence to the "situation": according to Halliday (1978), a situation consists of social action, role structure, and symbolic organization (cf. Halliday 1978: 142–45). The social action is referred to as "field" and is associated with the experiential, the role structure is referred to as tenor and is associated with the interpersonal, and the symbolic organization is referred to as mode and is associated with the textual. The symbolic organization is "the particular status that is assigned to the text within the situation; its function in relation to the social action and the role structure, including the channel or medium, and the rhetorical mode" (Halliday 1978: 141). Out of these aspects, "mode" has made an impressive career as a term.

When looking at frameworks for multimodal analysis, the influence of SFL is not always as opaque as in the early works of Kress and van Leeuwen. In *Reading Images*, Kress and van Leeuwen (1996/1998[3]) outline the development of their new interpretation of SFL as an advancement from language as social semiotic (Halliday 1978) toward social semiotics as a framework *not primarily* for linguistic analysis but for the visual realm, that is, images. There, the relation of mode and semiotic resources is of great importance. First, "mode" is understood as being dependent on the respective community: "Socially, a mode is what a community takes to be a mode and demonstrates that in its practices; it is a matter for a community and its representational needs" (Kress 2014: 65). These modes draw on semiotic resources (cf. Kress 2014: 61).

> Semiotic resources are the actions, materials and artefacts we use for communicative purposes, whether produced physiologically—for example, with our vocal apparatus, the muscles we use to make facial expressions and gestures—or technologically—for example, with pen and ink, or computer hardware and

> software—together with the ways in which these resources can be organized. Semiotic resources have a meaning potential, based on their past uses, and a set of affordances based on their possible uses. (van Leeuwen 2005: 285)

This quotation underlines that there is an interrelation between modes and semiotic resources, that is, categories for an approach that encompasses all aspects of communication without applying only linguistic categories to visuals but promoting the development of new categories and new terminology.

In Multimodal Discourse Analysis, the influence of SFL is stronger, as language plays a more important role there, especially in the works of O'Halloran. The relation between mode and semiotic resource is different from social semiotics, as mode and semiotic resource are being used interchangeably. Instead, the category "modality" plays an important role: "Semiotic resource is used to describe the resources (or modes) (e.g. language, image, music, gesture and architecture), which integrate across sensory modalities (e.g. visual, auditory, tactile, olfactory, gustatory, kinestetic) in multimodal texts, discourses and events, collectively called multimodal phenomena" (O'Halloran 2011: 121, original emphases).

In Multimodal (Inter-)Action Analysis, mostly represented by Norris (cf. Norris 2004), the understanding of mode mostly follows Social Semiotics, as can be explicated with the example of language: "We can view language as just one mode of communication that is present among other modes, without language necessarily being primary" (Norris 2004: 24). The differences to Social Semiotics and Multimodal (Inter-)Action Analysis lie within the research focus: "Multimodal Interactional Analysts set out to understand and describe what is going on in a given interaction. We analyze what individuals express and react to in specific situations" (Norris 2004: 4). This can be understood as a strictly holistic approach to studying interactions.

A rather uncomplicated definition of "mode" is provided by Bateman (2008). In his corpus linguistics-influenced approach to multimodal documents, the genre and multimodality model, he departs from analyzing written text only, as written text used to be the central focal point of written documents. "But things have changed: nowadays that text is just one strand in a complex presentational form that seamlessly incorporates visual aspects 'around,' and sometimes even instead of, the text itself. We refer to all these diverse visual aspects as *modes* of information presentation" (Bateman 2008: 1; original emphasis). Thus, Bateman provides a very basal definition of mode in which written language is one mode among others. This definition is further developed by Bateman (2011: 19–25) into a more complex, more detailed definition of *semiotic modes* that incorporates a distinction of semiotic modes and semiotic resources, the respective material substrate of a semiotic mode, and ties to discourse semantics.

These rather different emphases on central categories lead to a slightly provocative thought: perhaps the question "What is mode?" (Kress 2014: 60) is still widely and controversially discussed because of the lack of a homogeneous theoretical background. So the question should rather be, "What is understood as mode in a specific context of study?"

Apart from the approaches pointed out above, the Interactional Linguistics' take on multimodality, represented by Deppermann (2013) and Schmitt (2013) among others, forms a scientific discourse of its own, and it is not tied to Anglo-American discourses on multimodality. As Interactional Linguistics draws on Conversation Analysis (CA), it is primarily interested in interactions between at least two participants, where language plays one role among other modalities and is connected to them. The approach is not as holistic as Norris's Multimodal (Inter-)Action Analysis, as language is still understood as the primary object of research notwithstanding the possibility that communication can be realized through the use of modes that are not linguistic. This leads to a categorical problem: There is no systematic theoretical understanding of "mode" as Deppermann points out: "The terms 'multimodal interaction' and a fortiori 'multimodality' were not originally coined within a CA-context. They have been used by different communities in very different ways for very diverse kinds of phenomena" (Deppermann 2013: 2). Therefore, he criticizes the term "multimodality": "'multimodality' is a label which is already worn out and has become most fuzzy by its use in various strands of semiotics, discourse and media analysis" (Deppermann 2013: 2). Despite the critique as a "fuzzy label," the term "multimodality" is used in Interactional Linguistic and additionally the term "mode" is used extensively, without systematic reflection on the theoretical implications of this incorporation. Moreover, Interactional Linguistics seems to be more interested in the question of how multimodal phenomena modify classic categories of CA like simultaneity or sequentiality or how they make them redundant (see Deppermann, Schmitt and Mondada 2010: 1701). Thus, multimodality is likewise understood as a test for the theoretical and analytical framework.

The origins of CA's interest in multimodality are traced back to the works of Birdwhistell (1970) and Scheflen (1972), who "created an interdisciplinary interest in kinesics, gestures and verbal communication" (Deppermann, Schmitt and Mondada 2010: 1701) and the works of Goffman (1983), whose declared aim was "to study the coordination of action through multimodal resources and not only to focus on the verbal production of the participants" (Deppermann, Schmitt and Mondada 2010: 1701). It has to be pointed out that Goffman himself did not speak of "multimodal resources," and it is an important aspect of the comparison of Interactional Linguistics and other approaches to multimodality that the former discards the differentiation between mode and semiotic resources. This is a direct consequence of the nonconceptual use of the terms "multimodality" and "mode." Instead, in Interactional Linguistics, multimodality is more of a denomination for phenomena that have been analyzed under a different denomination (e.g., as nonverbal communication) and which arose anew—or at least gained importance—with new technological possibilities. The extension of technological possibilities not only affects everyday life but research methods as well. Thus, the turn toward multimodality within Interactional Linguistics has crucially been influenced by the use of video technology for data recording (cf. Deppermann 2013: 2).

3 Functional pragmatics

In the following paragraph, the aforementioned approaches to multimodality and multimodal analysis will be put into perspective with a linguistic approach, namely FP, that forms a scientific discourse on its own, more or less apart from the aforementioned approaches. To what extent multimodal phenomena are included in FP and how it can extend research on multimodality will be pointed out.

Whereas Interactional Linguistics uses the term "multimodality" to a certain extent, the perspective of FP toward it is different. In fact, close to no one in FP even uses the term "multimodality," and, thus, there is no such thing as a discourse on multimodality within FP, even though analyses within the framework would fit into that field of research. This has its reasons at the core of the theoretical framework that relies on highly critical analyses of categories and terms in general, as will be pointed out below. Although there is an analysis conducted by Bührig (2004), "On the multimodality of interpreting in medical briefings," an extensive discussion of the term "multimodality" is still to be had. In this analysis, Bührig (2004) shows that the use of diagrams to explain the medical treatment to be undertaken is employed by the doctors to impart medical knowledge (see Bührig 2004: 231–32), which leads to complex challenges for interpreters. For this purpose, Bührig includes the diagrams into her analysis, thus showing the interrelation of verbal actions and nonverbal actions and the importance of considering every aspect of a speech situation.

The dichotomy of nonverbal and verbal actions is of great importance: the term "nonverbal" is understood as a denomination for all communicative actions that are not verbal (cf. Ehlich 2013). The term "nonverbal" has been criticized in various discussions outside of FP: Fricke follows Kendon (1980) by opposing the assumption that gestures are nonverbal, as she shows that there are certain gestures accompanying speech that do have grammatical capacity (cf. Fricke 2015). In this chapter, the expression "nonverbal" is used in a different sense than Fricke's, following the distinction by Ehlich (2013). Ehlich proposes a systematization of nonverbal communication by defining that nonverbal communication forms a subsystem within human communication (cf. Ehlich 2013: 652). He categorizes different types of nonverbal communication: "Probably the most frequent form of nonverbal communication is concomitant, i.e. nonverbal communication accompanies verbal communication. In contrast to this, there is independent nonverbal communication, i.e. nonverbal communication that does not rely on or support verbal communication in order to be understood" (Ehlich 2013: 652, original emphases).

Thus, the categorization incorporates the relation of nonverbal and verbal communication systematically (cf. Graßer and Redder 2011). By approaching nonverbal and verbal communication this way, the communicative quality of gestures is not in dispute, and it is obvious that the analysis of modes and semiotic resources (in a holistic understanding that does not primarily focus on language) and their (inter-)relations is not the focal point of functional-pragmatic analyses.

Apart from the aforementioned and other functional-pragmatic analyses that take nonverbal communication into account (see an overview in Ehlich 2013: 649), there are a few functional-pragmatic analyses that affect the empirical scope of this chapter—multimodality in academic teaching and learning—even more directly. These are an analysis of academic knowledge dissemination in engineering lectures in pre-PowerPoint times (Hanna 2003); an analysis of pupils' presentations with PowerPoint in schools (Berkemeier 2006); analyses of students' note-taking (Redder 2009 and Breitsprecher 2010); and a recent analysis of the use of visual media, namely overhead projectors and PowerPoint, in economics lectures (Brinkschulte 2015). All of these analyses do not intend to approach multimodality as a field of research as such, but they widen the focus of linguistic analysis of knowledge dissemination by providing insights into the use of visual media in teaching and learning constellations.

Hanna and Brinkschulte show how pieces of knowledge are seemingly duplicated in the respective types of visual media and the oral verbalization. This could be misunderstood as reading out loud what is already readable on the projection. While there are undoubtedly situations in which lecturers read from the projection verbatim, there are a lot of situations in which they do not. Moreover, lecturers most often say more than what is projected: even by emphasizing several words from the projection, a lecturer adds something to writing on a projection. Besides, presentations have been prepared in previous speech situations with a supportive or additional character in mind. Visualizations, however, play a central role in a lot of disciplines and do contain condensed pieces of knowledge (cf. Hanna 2003). Thus, speech actions such as explaining, reasoning, and summarizing can become crucial for the successful transfer of knowledge. The challenge for an action-centered analysis of knowledge dissemination is that all of these aspects have to be taken into account, as a projection can consist of verbal and graphical elements deployed simultaneously. The analyses mentioned above point out that none of these aspects should be approached independently of each other. Thus, these analyses could be understood as multimodal research in a broader sense. To be more specific, these phenomena have been parts of action patterns from the very beginning of FP. Hence, calling the analyses "multimodal" would blur the theoretical localization of these works unnecessarily. This, however, needs to be backed up with a closer look into language theory.

FP is mostly a German-bound theory, as most researchers who work in FP have been socialized scientifically in German-speaking countries. Until now the theory has been exported to the Netherlands, Austria, Switzerland, Italy, Greece, Turkey, and Egypt. Thankfully, this issue was addressed by the development of a glossary (Ehlich et al. 2006)[4]. Apart from this, there are several handbook entries on FP in German (see Rehbein 2001 and Rehbein and Kameyama 2004) as well as introductory papers, such as Ehlich (1999 and 2010). Additionally, there is a handbook entry in English (Redder 2008), to which the author refers for a much more detailed account on FP.

FP is not a different take on pragmatics but a complete theory: "Pragmatics is not conceived of as 'pragmalinguistics', i.e., as a language-in-use-module to be added to language as a semiotic system.... On the contrary, Functional Pragmatics attempts to reconstruct language as an—abstract—societal action form through an analysis of authentic linguistic interaction" (Redder 2008: 134). The theoretical origins of FP lie within the speech-act theory of Austin (1962) and the theory of language by Bühler (1990). "Functional Pragmatics expands Bühler's *organon*-concept and employs *purpose* as a central category in a twofold manner: *extralinguistic purposes*, as they are pursued by societal actants, and *linguistic purposes*, according to which language itself is shaped" (Redder 2008: 134; emphases in original). One of the most important aspects of FP is that language is systematically understood as an action between at least two participants and therefore—as will be pointed out in more detail—the speakers' as well as the hearers' mental processes are integral parts of the analysis. In short, FP is an action theory of language that specifically investigates the relation of society and individual.

> Society is the sociohistorical base category from which the category of the individual is derived (not vice versa). Individuals, as societal actants, pursue purposes, i.e., repetitive societal needs to be satisfied through actions. . . . Actants' personal goals are thus always structurally related to purposes. As reality is societal reality, actants' needs for action arise in repetitive constellations. Such a constellation, for instance a knowledge deficit, may be dealt with through linguistic action. The paths for such actions are societally elaborated as linguistic action patterns. (Redder 2008: 135; emphasis in original)

These actants are speaker (S) and hearer (H)—both to be understood as abstract categories that do not limit an actant to that role. These patterns are acquired during the speakers' socialization as well as the "pattern knowledge" (cf. Ehlich and Rehbein 1977: 66–68). Rehbein elaborates: "These patterns are composed of different types of actions, namely mental actions, interactions, and non-communicative actions, and these three must again be differentiated into procedures, acts, and actions. The main point is that empirical analysis of action and speech . . . has to take into considerations various kinds of mental procedures, acts, and actions" (Rehbein 1984: 62).

Linguistic action in this understanding "involves three dimensions of reality: extralinguistic reality (capital letter P), for instance, the constellation that gives rise to the speaker's knowledge deficit; mental reality (Π-area) within which the speaker determines his precise non-knowledge so that he can ask a question; and linguistic reality (small letter p), i.e., the speaker's linguistic action, the question" (Redder 2008: 136; emphases in original).

According to Ehlich and Rehbein (1986), the relation of the three dimensions of reality, and thus the structure of knowledge, can be modeled as shown in Figure 6.1.

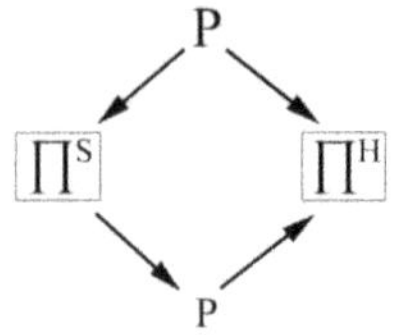

Figure 6.1 Basic model of speech actions (Ehlich and Rehbein 1986: 96).

Thus, the "success of speech actions requires a *synchronization of the speaker's and hearer's Π-areas* with respect to topics, focus of attention, previous (speech-) actions, etc. To achieve this synchronization, there are language-specific devices" (Redder 2008: 138; original emphasis). These devices can be speech actions like apology, justification, or substantiation, smaller linguistic devices, called "procedures" (cf. Redder 2008: 139–40), that are not necessarily identical to parts of speech. These units of actions can be tied to, combined with, or substituted by, nonverbal actions.

As it is a basic assumption of FP that the rather abstracts aspects of interaction—the synchronization of mental realities, for example—are reconstructible through linguistic analysis, the corollary for this chapter is the question of if and how these aspects can be realized as multimodal. This form of linguistic analysis only works through the analysis of authentic, empirical data, as FP strongly opposes the analysis of constructed, artificial data. Rehbein is

> pleading for an analytic approach which starts from concrete instances of spoken or written language and aims at gaining insight into the regularities of social life which express themselves in such occurrences of speaking and writing . . . . This analysis has . . . to start from recorded or written instances of actual communication by which social processes are realized. (Rehbein 1984: 40)

This is laid out in great detail in the development of a system for the transcription of spoken language, HIAT (cf. Ehlich 1993), which is introduced in the following paragraph (for a summary of analytical steps necessary for an FP analysis, see Redder 2008: 142–43). Through this approach, the societal aspect of all actions is assumed to be reconstructible from authentic actions processed productively (by S) as well as receptively (by H). This reconstruction has been conducted most notably in analyses of language in institutions[5] like schools (Rehbein 1984), hospitals (Bührig and Meyer 2014), or universities (Thielmann, Heller and Redder 2015). In institutions, characteristic modifications concerning the actants are of great importance: "*Agents* (i.e., actants who act on behalf of an institution) realize institutional purposes and usually act on the basis of second degree institutional knowledge, i.e., an institutional theory such as educational sciences provide for the institution school. Clients, on the other hand, are actants who avail themselves of institutional purposes for their individual goals" (Redder 2008: 145–46; original emphases).

Through the analysis of knowledge transfer in institutions, especially in hospitals, schools, and universities, problematic communicative processes can be reconstructed that are crucial for the institution. One prominent aspect for all three mentioned institutions are multilingual challenges, connected to nearly every field in the institutions and affecting both agents and clients (for analyses of linguistic landscaping from an FP perspective, see Redder 2013). "In short, the fundamental aim of Functional Pragmatics is to analyze language as a sociohistorically developed action form that mediates between a speaker (S) and a hearer (H), and achieves—with respect to constellations in the actants' action space . . . —a transformation of deficiency into sufficiency with respect to a system of societally elaborated needs" (Redder 2008: 136).

The biggest challenge for FP is and will be the interchange with other theories. For example, Januschek, Redder, and Reisigl (2012) point out that ties between different linguistic theories like FP, SFL, and Critical Discourse Analysis (CDA) (in its varieties) can be found and should be looked into further. While Reisigl (2012) looks into ties between FP and CDA on a theoretical level, Gruber (2012) displays ties between SFL and FP and points out that one important difference is the respective reception of Bühler. While SFL is a semiotic theory that primarily focuses on the *organon*-model, FP systematically works with Bühler's linguistic fields and develops the idea further.

From this point of view, the combination of FP and multimodality is not an easy task. This, however, depends on the respective form and understanding of multimodality. Whereas it has been pointed out that one origin of multimodal research lies within SFL as a transition of Halliday's ideas, the numerous perspectives on multimodality nowadays suggest the assumption that multimodality has become an umbrella term for vastly heterogeneous research from many different disciplines and theoretical approaches. Incorporating "multimodality" (and thus "mode") as a category into FP would blur the theoretical background of FP, given that at least some objects of multimodal research are not new to FP research but have been analyzed under different denominations. Hence, multimodality could function as a non-categorical expression for interdisciplinary mutual understanding. Keeping this in mind, a functional-pragmatic perspective on multimodality would be linguistic analysis that systematically considers multimodal resources as potentially integral aspect of interactions and communication.

Setting aside the theoretical challenges of the term "multimodality" itself, what are the benefits of an action-centered approach like FP for the analysis of multimodal communication? Where are the differences to approaches that specifically look at data from a multimodal perspective? This can be answered tentatively by looking at Iedema's (2001) idea of resemiotization as applied to multimodal data in academic settings as conducted by Fortanet Gómez and Crawford Camiciottolo. They provide the example of a conference presentation to show how the outcome and the planning change semiotically:

> Planning is organized very often in writing, and later the written document becomes an oral performance (usually embedding a number of semiotic modes in itself), which turns again into a written text when it is published in the proceeding of a conference or any other printed version of the paper or lecture presentation. In each of these changes, the communicative event undergoes resemiotization since the modes and their relationships are different. (Fortanet Gómez and Crawford Camiciottolo 2015b: 3)

The semiotic resources and modes used in the process obviously do change and do influence each other. However, one question remains unanswered: *why* the different semiotic resources and modes are being used in the respective situations.

From a purpose-oriented point of view, this is rather easy to answer: At every stage of the process of planning a conference presentation, the used medium or form of communication is a reaction to the respective tasks. In the stage of planning, written text is used to bridge the temporal gap between the speech situations in which the planning is conducted, as the planning of a conference presentation normally extends over several hours or days. Additionally, the written text can be used later as the basis for the oral presentation, which is usually accompanied by slides or similar media aids. Thus, spoken language and other "modes," like gestures or kinesics, are employed. The purpose of PowerPoint presentations (or any other type of visual media) is rather difficult to determine without concrete data and is currently under investigation (see for example Krause 2015a for an action-centered approach and Diani 2015 for a multimodal approach)—tentatively, the purpose is to support the process of the recipients' processing of knowledge by making central points available in written and spoken language at the same time. Additionally, graphics, charts, and the like can visualize abstract and complex matters that cannot be verbalized as readily in spoken language. This characterization is only tentative and needs much more explication than can be accomplished in this chapter.

In this sense, it is not surprising that different semiotic resources and modes are being used. Compared to describing "the modes the speakers can choose from, how speakers choose different modal resources in a communicative event, and how speakers interact with each other by negotiating meaning in certain context such as scientific conferences and university classes" (Fortanet Gómez and Crawford Camiciottolo 2015b: 5) as mostly conducted by the contributions in Fortanet Gómez and Crawford Camiciottolo (2015a), an analytical focus on the respective purpose of the actions in the communicative events can lead to different findings.

This brief comparison of the two approaches only scratches the surface of the capability of functional-pragmatic analyses of multimodal data. As a fully developed linguistic theory, analyses of the micro level of teaching and learning constellations could provide insights into the organization of knowledge dissemination at the atomic level of language and in systematic interconnection to nonverbal communication and respective types of visual media.

4 FP methodology for multimodal research

As the theoretical thoughts on the purpose of visual media in academic teaching and learning above have shown, the question of purpose cannot be addressed without the analysis of concrete data. Apart from the aforementioned analytical focus within FP, there is also a longtime experience of transcribing data with a fully developed methodology (for an overview of different transcription systems, see Redder 2001). This methodology can complement the analysis of multimodal data from a linguistic perspective—at least when spoken language is involved. Spoken language itself brings specific challenges with it: "Oral data themselves are evanescent. Preserving them is a first and primary objective for all further steps of analysis. If data are not restricted to an abstract language system, it is also necessary . . . to account for phenomena that have to do with the interactive character of communication between speakers and listeners" (Ehlich 1993: 124).

This can be conveyed through the use of transcripts. From the very beginning of FP, the transcriptions did not only focus on verbal activities "but also [on] paralinguistic, nonverbal, and, partly, actional activities. These have to be accounted for in the process of making the evanescent communication events permanent and accessible to further analysis" (Ehlich 1993: 124). For this purpose, Ehlich and Rehbein (1976 and 1979) developed a system for transcription, the HIAT system; as Ehlich points out,

> the HIAT system . . . was developed with three criteria in mind: (a) simplicity and validity, (b) good readability and correctability, and (c) minimum of transcriber and user training. The acronym, HIAT, stands for *Halbinterpretative Arbeitstranskriptionen* [German for "Semi-Interpretative Working Transcriptions"]. "Interpretative" refers to the overall hermeneutic process of understanding the spoken data. That the process is open to further analytical steps is reflected in the qualification of the name as being "semi-interpretative" (*halbinterpretativ*). (Ehlich 1993: 125; original emphases)

Given these criteria, HIAT does not rely upon the alphabetic system of phonetic notation provided by the International Phonetic Alphabet (IPA). "The HIAT system uses, instead, a derivation from written orthography which we call *literary transcription* or in German, *literarische Umschrift*. . . . Literary transcription involves systematic departures from the standard orthographic rendering of an item but in a manner that is meaningful to someone familiar with the orthographical system as a whole" (Ehlich 1993: 126; original emphases).

Thus, readability is one of the leading principles when transcribing in HIAT—while obtaining important characteristics of everyday spoken language such as pauses, interruption, tones, modulation, or parallel speaking by transcribing those characteristics. For this purpose, a system of signs is needed, as Rehbein et al. (2004) point out for the transcription with the EXAMRaLDA Partitur-Editor (for a step-by-step-tutorial for using the Partitur-Editor in English, see Rehbein 2008).

An outstanding accomplishment in HIAT is the transcription of the simultaneity of speech and action of more than one speaker, as spoken language rarely occurs in monologue form. As Ehlich points out,

> Standard Greek/Latin-based writing systems are inadequate for this purpose because the flow of time is represented from left to right, only one line at a time. In transcribing discourse data . . . it is desirable to expand the dimension of a transcript to allow for several simultaneous, ongoing events, in a graphically neat, straightforward manner. A highly effective solution is suggested by the system used in musical representations—*the musical score*. (Ehlich 1993: 129; original emphasis)

By this form of representation, the simultaneity of spoken language is documentable and readable. The score notation of linguistic actions can be extended to nonverbal actions: the simultaneity of nonverbal actions and verbal actions can be documented in the same way as parallel speaking. In the same ways that every speaker is assigned to a unique tier, for every speaker additional tiers can be designated for the modes used in the respective interaction. Depending on the focus of the analysis, it might be a suitable solution to assign only one tier for all kinds of nonverbal actions to each speaker. If several aspects of nonverbal actions are to be analyzed separately, a separate tier for every analyzed body part might be the best solution. This makes HIAT the perfect solution for a linguistic approach toward multimodality, that is, data that consider which spoken language is predominant—especially with the extensions to HIAT proposed in this chapter (see below).

From a methodological point of view, FP can draw from a lot of experience transcribing audio and video data of interactions. Comparing the approach of HIAT in FP with Norris' Multimodal (Inter-)Action Analysis approach to transcription (cf. Norris 2004), it can be said that both the approaches could benefit from each other. Multimodal (Inter-)Action Analysis has its benefits in its strictly holistic approach to interactions, with language systematically not being the prime focus. HIAT, on the other hand, provides readability, accessibility, and a fully developed system for the transcription of verbal, nonverbal, and actional elements of interactions. In short, interdisciplinary work needs to be done to evaluate the possibilities to bring these approaches closer together.

5 Multimodality in academic teaching and learning

On the basis of the theoretical and methodological groundwork pointed out above, it will be laid out in the following, how HIAT can be expanded for the transcription of multimodal data. This will be explicated using an example from a tutorial on mathematics, taken from a research project on academic teaching and learning, the euroWiss-project.[6] In the context of this explication, a step-by-step proposal for a digital reconstruction of writing on a blackboard in relation to the spoken language is provided.

As pointed out above, FP is a theory that allows for linguistic reconstruction of knowledge, as mental processes are systematically understood as objects of research that are reconstructible through the analysis of linguistic actions. In institutions that execute the processing of knowledge from one generation to the next, such as schools or universities, the analysis of knowledge becomes crucial to identify potential communicative challenges, problems, or developments. For example, this can be the challenge of multilingual universities and a European academic system that—at least hypothetically—encourages exchange between universities in different countries and different academic cultures.

Academic teaching and learning nowadays strongly rely upon the use of visual media in every conceivable way. However, it would be a great mistake to assume that the use of visuals arose in the last ten or fifteen years, through the popularity of PowerPoint. While PowerPoint is increasingly berated or campaigned against in the press (mostly associated with the so-called "PowerPoint is evil" debate, led by Edward Tufte, which had its beginnings in Tufte 2003), there has been a type of visual medium before PowerPoint that is overlooked all too often in media studies and in linguistics alike: the blackboard.

Keeping this in mind, academic teaching and learning has been multimodal long ago. Apart from it "only" consisting of spoken language, gesture, intonation, and so on, it also counts for different types of visual media. Based on observations from the aforementioned euroWiss-corpus, it could be said that several types of visual media are in use nowadays: blackboards, whiteboards, overhead projectors, PowerPoint, and interactive whiteboards, as well as digital or printed texts and other media. Additionally, the data in the euroWiss-corpus have shown that close to all German university courses use at least one type of visual media. Moreover, in most courses two or more different types of visual media are employed. For the lecturers, this leads to the challenge of having to orchestrate the different types of media to achieve knowledge dissemination. At the same time the students should not only be able to follow the knowledge dissemination across all types of visual media but also take notes. These notes also have to adapt to the types of visual media used in the courses: Breitsprecher's (2010) and Redder's (2009) analyses of authentic notes taken by students in lectures on German linguistics have shown that students develop individual semiotic systems within their notes. For example, students used arrows or different colors to visualize concatenations of propositional contents. By aligning the notes to the linguistic data in the respective lectures, Breitsprecher reconstructed the notes as a "window" to the individual mental processing of knowledge (cf. Breitsprecher 2010: 201).

Given the variety of visual media used in parallel, it is obvious that different types of visual media might realize different goals. This implies that an analysis of academic teaching and learning has to consider the interplay of verbal actions, nonverbal actions, and visual media—an analysis that does not consider these aspects misses out on a central characteristic of academic teaching and learning. As other analyses have already shown (cf. Krause 2015a), this counts for the natural sciences as well as for the humanities. This observation leads to the necessity of transcribing visual media or at least an addendum to the transcription of verbal and nonverbal actions. This will be explicated further using an example from a tutorial on mathematics from the euroWiss-corpus.[7]

In mathematics in general the blackboard plays a central role in knowledge dissemination and in research, as Greiffenhagen (2014) has pointed out. Greiffenhagen has found that mathematicians even fought for "their" blackboards if university administrations tried to replace those with whiteboards or interactive whiteboards (cf. Greiffenhagen 2014). The mathematical courses in the euroWiss-corpus underline the preference for blackboards over other types of visual media. This significance of the blackboard has direct consequences for the empirical analysis: as writing on the blackboard is handwritten, the usage of screenshots for the publication of an analysis can lead to troubles reading it. Additionally, the emergence of the writing on the blackboard in coherence to the spoken language is of great importance. Thus, to avoid inserting miniature screenshots, several analytical steps are necessary.

First, the writing on the blackboard has to be copied from the video recording by hand. As the handwritten copy provides the basis for the next steps, it is of great importance that every word is double-checked and that every spelling mistake is corrected and every abbreviation on the blackboard is meticulously copied. In the next step, the whole writing on the blackboard is split into segments according to its emergence parallel to the spoken language. These segmentations of the writing on the blackboard are not to be understood as units of actions but as methodological concessions to the attempt to make writing on the blackboard accessible for the incorporation into a HIAT transcription. Hence, following the HIAT conventions, the overall goal of the segmentation of the writing on the blackboard is to maintain readability.

In many cases, the segmentation of the writing on the blackboard and the mathematical operations were identical. But lecturers tended to pause at some points, for example to explain the operation or let students solve a mathematical problem. In those cases, the segments had to be fragmented according to their emergence. Whereas the segments usually were numbered chronologically with natural numbers, the fragmented segments were additionally numbered with letters. The decision over to what extent the segments should be fragmented cannot be generalized. Segmentation following the content has proved to be the most suitable strategy; at the same time, readability is warranted. The reason for this segmentation is that the numbers can be included in the HIAT transcript and be aligned with the spoken language (see Figure 6.2).

[39]

	/70/	/71/	
D [v]	(Nun Frage): Was setzen mer u Strich, was setzen mer v, Sm4 ?	((2,3s))	Von dem
D [engl]	(Now question): What do we set u line, what do we set v, Sm4?		Of this
D [nv]			*zeigt*
D [nv]			*points*
[nn]		((Räuspern))	
[engl]		((clearing thorat))	

[40]

			/72/
D [v]	______	kennen wir die Stammfunktion, deswegen ham mers so zerlegt.	Also was wird es?
D [engl]		we know the antiderivative, that's why we disassembled it that way.	Thus what will it
D [nv]	*auf 4b)*		
D [nv]	*at 4b)*		

[41]

		/73/	/75/	/76/	/77/
D [v]		((1,3s))	(Wenn mer die)/ U Strich, ne?	((2,3s))	So. Wenn mer
D [engl]	become?		(If we)/ U line, right?		So. If we want to,
D [nv]				*schreibt 5b) an die Tafel*	*schreibt 5c) an die*
D [nv]				*writes 5b) on the blackboard*	*writes 5c) on the*
			/74/		
Sm4 [v]			U Strich.		
Sm4 [engl]			U li ne.		

[42]

		/78/
D [v]	wollen, machen mer das wieder ganz ausführlich.	• • • Wir stopfen also in dieses
D [engl]	we do this very detailed.	So we cram into this equal sign,
D [nv]	*Tafel*	*schreibt 5d) an die Tafel*
D [nv]	*blackboard*	*writes 5d) on the blackboard*

Figure 6.2 Transcription of a tutorial on mathematics (excerpt).

In the next step, the handwritten copy is digitally reconstructed in Microsoft Word. This might, admittedly, be realized more professionally with the use of LaTeX, but so far Word has proved to be sufficient. This digital reconstruction is the basis for the compilation of the reconstructed writing on the blackboard. Every single segment is put into quadrats with dotted lines to signify that the lines are not part of the writing on the blackboard. The result is an organized, readable, and easy-to-understand segmentation that enables the reader to follow the emergence of the writing on the blackboard within the HIAT transcription and, thus, to align the writing on the blackboard with the respective spoken language in the knowledge dissemination (see Figure 6.3).

The following transcription of a mathematical tutorial is taken from the aforementioned euroWiss-corpus. The tutorial is led by a female professor who is very popular among the students, of whom nine were present at the time of recording. The transcription was conducted for the purpose of analyzing where the lecturer hesitates in the problem-solving, and, more specifically, which parts of the mathematical operations are handled by the lecturer and which parts are given as tasks to the students. Additionally, the writing on the blackboard was reconstructed, to understand which parts of the task are written down and on which points the lecturer reactivates former mathematical operations that

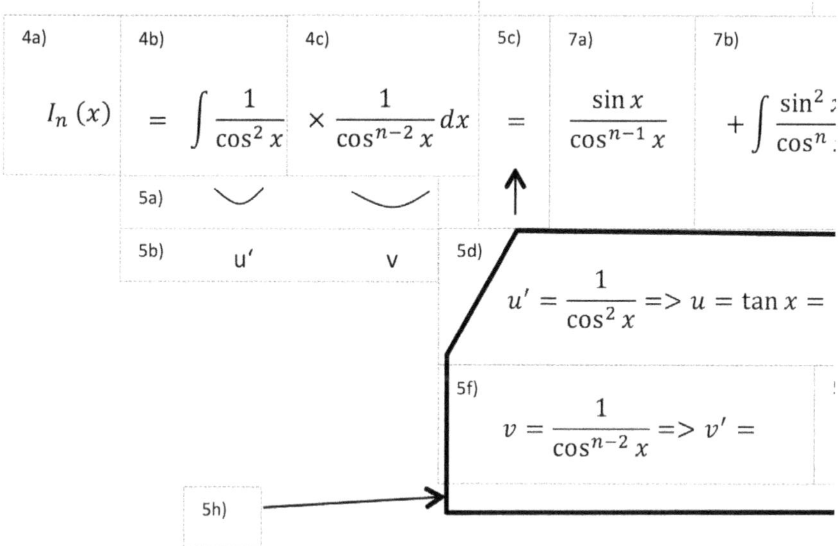

Figure 6.3 Small excerpt of the writing on the blackboard (digital reconstruction, cut off).

are crucial for the task solving. Thus, the focus lies upon the spoken language, the writing on the blackboard, and a few nonverbal aspects, such as pointing, without systematically transcribing all of them.

Emphases are labeled with underlines, the slash (/) symbolizes repair. The passages in brackets are not well audible, so that after several cycles, the transcribed words are the most suitable approximation. The double brackets with numbers are pauses, as well as the big black dots, each of which symbolizes a pause of 0.25 seconds. As this is an excerpt, all utterances are numbered beginning at seventy—the utterances have been numbered to make references to utterances easier.

The lecturer (D) in the following transcription is solving a mathematical task with the students. In this brief excerpt, only one male student (Sm4) is speaking. Unidentifiable speakers are marked as [nn], nv stands for "nonverbal," whereas "v" stands for verbal according to Rehbein et al. (2004: 74).

This short excerpt is an example of a traversal of an action pattern, here the "task completion pattern" (cf. Ehlich and Rehbein 1986). It has mostly been reconstructed in school, due to its central significance there. Whereas it can be expected that there is a qualitative change from the instructional teaching in schools to academic teaching and learning in universities (cf. Ehlich 1981), one would not necessarily expect to find the same action patterns (cf. Krause 2015b).

In this excerpt, the equation has to be solved to proceed from 4b and 4c to 7a and 7b (see Figure 6.3). This requires certain auxiliary calculations by students to disassemble parts of the equation, so that basal knowledge of antiderivatives can be utilized for the problem-solving. This can be achieved through conversions of the equation. The auxiliary calculations can be understood as utilizations of mathematical rules that provide the possibility to proceed toward solving the problem.

The transcript shows that D formulates the task in utterance /70/, in which Sm4 is picked to solve the problem. The following utterances in /71/ and /72/ are consequences of the relatively long pause of 2.3 seconds and thus the lack of an answer by Sm4: As Sm4 remains silent, D has to anticipate which pieces of knowledge are missing for him to solve the problem. Functionally seen, the utterance in /71/ is a hint that is the result of the lack of an answer. Thus, D wants to help Sm4 find the answer by providing additional information. The second part of /71/ ("deswegen ham mers so zerlegt"; "that's why we disassembled it that way") functions as a reasoning (cf. Ehlich and Rehbein 1986) for the previous mathematical operations. This mentally reactivates the prehistory of the operations.

In addition to the verbal hint in /71/ in general, D points physically and verbally at a specific part of the mathematical equation (segment 4b), and by that it refers to the crucial pieces of knowledge that all students are supposed to possess, that is, the antiderivatives that are required for the auxiliary calculation. Through this reference, Sm4 knows which parts of his knowledge he needs to solve the given task. Through this complex action, all of the relevant pieces of knowledge are supposedly available in the Π-area of Sm4. At the same time, D begins to give Sm4 another hint (see utterance /75/), but that utterance is discontinued and instead Sm4's answer is appraised.

Hence, the task formulated in /72/ is seemingly not a completely new task but more or less the same task as in /70/, only this time it is incorporating the hint in /71/. The reference to the hint is included in the German "also" ("thus"), which realizes an operative procedure with a deictic share: "also" helps us realize a "qualitative turn from the requirement of advance to drawing a conclusion" (translated from Redder 1989: 403)—in this case, pointing at the hint and henceforth incorporating it into the aforementioned task. At the same time, the task is modified, due to its difficulty: The task itself is relatively easy, as Sm4 only has to identify what has to become u' and, thus, the other part of the equation can be set as v—polemically expressed, a fifty-fifty chance. So if Sm4 did not know the answer, he could have guessed. Given the audio and visual data of Sm4 and D, it does not seem probable that he is guessing. Through the combination of the physical pointing at 4b in /71/, the question is no longer about which part of the equation is set *u'* and which part of it is *v*, but about what 4b has to become (referred to by "es" ("it") in/72/). Thus, the very brief answer of the student in/74/is fully sufficient and is written out by D in 5b.

To summarize, academic teaching and learning are multimodal. If the transcription above had not incorporated the writing on the blackboard and the gestures of the lecturer, an analysis could not have been conducted properly. Especially, the action of verbal and nonverbal pointing in/71/could not have been understood in the same way—what "es" or "it" in/72/refers to would not have been easy to reconstruct. It is primarily through the action of pointing to specific elements on the blackboard that the task is modified. Thus, the linguistic analysis of academic teaching and learning needs to include multimodal aspects to enable a reconstruction of the knowledge dissemination to the full extent. For this purpose, the analytical tools have to be developed further and have to adapt to the possibilities the use of video provides. In FP, the analysts can resort to many past experiences since video transcription has been systematically considered from the earliest stages of development of HIAT. This provides the basis for an adaptation of the system to new types of media that are being used, as has been shown in this chapter.

6 Conclusion

This chapter made the attempt to show a functional-pragmatic perspective to multimodality, that is, a linguistic, action-centered, and purpose-oriented perspective. It was pointed out that, even though multimodal phenomena were incorporated into analyses from the very beginning of FP, ties to the existing approaches toward multimodality have not been addressed as of yet. For this purpose, various approaches to multimodality were briefly described to point out that the different understandings of central categories lead to certain challenges in incorporating multimodality into FP and into other approaches. For example in Interactional Linguistics, "multimodality" and "mode" are not used as concrete terms but merely as umbrella terms for phenomena that have previously not been the object of analysis.

To expand the research landscape of multimodality further, this chapter sought to introduce a different approach to multimodality. For this purpose, an overview of FP as an action-centered linguistic theory and its possible connecting points to multimodal research were presented. It was demonstrated that the purpose orientation of FP leads to insights into multimodal data with regard to mental processes and the structure of knowledge as these are, systematically, objects of analysis. They are accessible through the analysis of verbal actions in relation to nonverbal actions and visual media. It was pointed out in this context that the HIAT system developed for the transcription of spoken language—with the extensions pointed out in this chapter—can be a powerful tool for the transcription of multimodal data and thus provide the basis for a multimodal analysis with language as a starting point. By this, a language-centric multimodal analysis allows to reconstruct mental processes across many "modes." FP provides the theoretical background for this. Also it was highlighted that there are interesting ties to Multimodal (Inter-)Action Analysis. These ties should be studied in much more detail than could be tackled in this chapter—from both a theoretical and a methodological perspective.

As shown above, the analysis of knowledge dissemination in academic teaching and learning can be conducted in a very productive way with the help of functional-pragmatic theory and methodology: In the concluding sample analysis of a mathematical tutorial taken from the euroWiss-project, it could be shown that academic teaching and learning rely on an interplay of spoken language, visual media, and nonverbal communications. This interplay is analyzable through the use of functional-pragmatic theory and methodology with regard to the mental processes of the actants.

To conclude, bringing an FP perspective into the scientific discourse on multimodality could strengthen discussions on how knowledge is structured—through all means of communication. And vice versa: bringing a multimodal perspective into FP can be challenging yet compelling, as more aspects of communication are to be considered systematically.

Notes

1 The author thanks Professor Dr. Angelika Redder for her helpful and critical commentaries on this chapter, which enormously helped specify crucial aspects. Additionally, the author thanks Susanne Krause for carefully commenting on grammatical issues in this chapter. The author takes responsibility for any remaining inaccuracies.

2 For other pragmatic theories and approaches, see, for example, Bublitz and Norrik (2011) and the subsequent editions in the series; these theories will not be laid out in this chapter.

3 *Reading Images* was published in 1996, but unfortunately at the point of writing, only a reprint from 1998 was available. As a concession to this unpleasant situation, the author chose to refer to it by citing it as 1996/98.

4 The languages currently covered in the glossary are German, English, and Dutch; extensions to other languages (Italian, Greek, and Turkish) are in progress.

5 Note: the publications mentioned below are not the only or the most important, or the most recent ones. It was the author's intention to list publications in English, primarily to make FP more accessible.

6 This has been the research focus of the project "euroWiss—Linguistic profiling of European Academic Education" funded by the VW Foundation between 2011 and 2014 (for the project's aims, see Heller, Hornung, Redder and Thielmann 2013; for key findings see Thielmann, Redder and Heller 2015). In the project, approximately 350 hours of university courses in Germany and Italy were audio and video recorded and compiled in the euroWiss-corpus. The recorded courses were situated within "a wide range of disciplines such as economics, sociology, linguistics, literature, mathematics, physics, and mech[a]nical engineering" (Thielmann, Redder and Heller 2015: 235). The possibility to access data from this wide range of disciplines allows to look into diverse ways of academic teaching: On the one hand, the Humboldtian ideal of a combination of research and academic teaching was observed in all of the courses recorded in Germany. On the other hand, first analyses of lectures from the humanities and natural sciences showed specific differences between the two faculties. When comparing these results to the Italian data, the difference between the two countries' introductory courses in the natural sciences turned out to be rather marginal (Krause and Carobbio 2014), whereas comparisons of humanities courses showed huge differences between entry-level and advanced courses (Carobbio and Zech 2013).

7 The author thanks Professor Dr. Winfried Thielmann for the permission to use the transcription from the euroWiss-corpus.

References

Austin, J. R. (1962), *How to Do Things With Words*, Oxford: Clarendon.

Bateman, J. A. (2008), *Multimodality and Genre: A Foundation for the Systematic Analysis of Multimodal Documents*, London: Palgrave Macmillan.

Bateman, J. A. (2011), "The Decomposability of Semiotic Modes," in K. L. O'Halloran and B. A. Smith (eds.), *Multimodal Studies: Multiple Approaches and Domains*, 17–38, London: Routledge.

Bateman, J. A. (2014), *Text and Image: A Critical Introduction to the Visual/Verbal Divide*, London: Routledge.

Bateman, J. A. and K.-H. Schmidt (2012), *Multimodal Film Analysis: How Films Mean*, London: Routledge.

Berkemeier, A. (2006), *Präsentieren und Moderieren im Deutschunterricht*, Hohengehren: Schneider Verlag.

Breitsprecher, C. (2010), "Studentische Mitschriften—Exemplarische Analysen, Transformationen ihrer Bedingungen und interkulturelle Forschungsdesiderate," in D. Heller (ed.), *Deutsch, Italienisch und andere Wissenschaftssprachen—Schnittstellen ihrer Analyse*, 201–16, Frankfurt: Peter Lang.

Birdwhistell, R. (1970), *Kinesics in Context*, Philadelphia: University of Pennsylvania Press.

Brinkschulte, M. (2015), *(Multi-)mediale Wissensübermittlung in Vorlesungen. Diskursanalytische Untersuchungen zur Wissensübermittlung am Beispiel der Wirtschaftswissenschaft*, Heidelberg: Synchron.

Bucher, H.-J. (2011), "Multimodales Verstehen oder Rezeption als Interaktion: Theoretische und empirische Grundlagen einer systematischen Analyse der Multimodalität," in H. Dieckmannshenke, M. Klemm, and H. Stöckl (eds.), *Bildlinguistik. Theorien—Methoden—Fallbeispiele*, 123–56, Berlin: Erich Schmidt Verlag.

Bublitz, W. and N. R. Norrik (eds.) (2011), *Foundations of Pragmatics: Handbooks of Pragmatics 1*, Berlin: de Gruyter.

Bühler, K. (1990), *The Theory of Language: The Representational Function of Languag*, trans. by Donald Fraser Goodwin, Amsterdam/Philadelphia: John Benjamins.

Bührig, K. (2004), "On the Multimodality of Interpreting in Medical Briefings," in E. Ventola, C. Charles, and M. Kaltenbacher (eds.), *Perspectives on Multimodality*, 227–47, Amsterdam/Philadelphia: John Benjamins.

Bührig, K. and B. Meyer (2014), "Interpreting Risks: Medical Complications in Interpreter-Mediated Doctor-Patient Communication," *European Journal of Applied Linguistics*, 2 (2): 233–53.

Carobbio, C. and C. Zech (2013), "Vorlesungen im Kontrast—Zur Vermittlung Fremder Literatur in Italien und Deutschland," in B. Hans-Bianchi, C. Miglia, D. Pirazzini, I. Vogt and L. Zenobi (eds.), *Fremdes ahrnehmen, aufnehmen, annehmen—Studien zur deutschen Sprache und Kultur in Kontaktsituationen*, 289–308, Frankfurt: Peter Lang.

Deppermann, A. (2013), "Multimodal Interaction from a Conversation Analytic Perspective," *Journal of Pragmatics*, 46: 1–7.

Deppermann, A., R. Schmitt, and L. Mondada (2010), "Agenda and Emergence: Contingent and Planned Activities in a Meeting," *Journal of Pragmatics*, 42: 1700–18.

Diani, G. (2015), "Visual Communication in Applied Linguistics Conference Presentations," in I. Fortanet Gómez and B. Crawford Camiciottoli (eds.), *Multimodal Analysis in Academic Settings. From Research to Teaching*, 83–107, London: Routledge.

Ehlich, K. (1981), "Schulischer Diskurs als Dialog?" in P. Schröder and H. Steger (eds.), *Dialogforschung*, 334–69, Düsseldorf: Schwann.

Ehlich, K. (1993), "HIAT: A Transcription System for Discourse Data," in J. A. Edwards and M. D. Lampert (eds.), *Talking Data. Transcription and Coding in Discourse Research*, 123–48, Hillsdale: Lawrence Erlbaum.

Ehlich, K. (1999), "Funktionale Pragmatik—Terme, Themen und Methoden," *Deutschunterricht in Japan*, 4: 4–24.

Ehlich, K. (2007), *Sprache und sprachliches Handeln*, Three Volumes. Berlin: de Gruyter.

Ehlich, K. (2010), "Sprechhandlungsanalyse," in L. Hoffmann (ed.), *Sprachwissenschaft: Ein Reader*, 3. Auflage, 242–57, Berlin: de Gruyter.

Ehlich, K. (2013), "Nonverbal Communication in a Functional Pragmatic Perspective," in C. Müller, A. Cienki, E. Fricke, S. Ladewig, D. McNeill, and S. Teßendorf (eds.), *Body—Language—Communication: An International Handbook on Multimodality in Human Interaction*, 648–58, Berlin: de Gruyter.

Ehlich, K. and J. Rehbein (1976), "Halbinterpretative Arbeitstranskriptionen (HIAT)," *Linguistische Berichte*, 45: 21–41.

Ehlich, K. and J. Rehbein (1977), "Wissen, kommunikatives Handeln und die Schule," in H. C. Goeppert (ed.), *Sprachverhalten im Unterricht*, 36–114, München: Fink.

Ehlich, K. and J. Rehbein (1979), "Erweiterte Halbinterpretative Arbeitstranskriptionen (HIAT 2): Intonation," *Linguistische Berichte*, 59: 51–75.

Ehlich, K. and J. Rehbein (1982), *Augenkommunikation: Methodenreflexion und Beispielanalyse*, Amsterdam: John Benjamins.

Ehlich, K. and J. Rehbein (1986), *Muster und Institution*, Tübingen: Narr.

Ehlich, K., L. Machenzie, J. Rehbein, W. Thielmann, and J. D. ten Thije (2006), A Glossary for Functional Pragmatics, March 31. Available online: http://www.jantenthije.eu/wp-content/uploads/2010/09/FP-Glossar-90506.pdf.

Fortanet Gómez, I. and B. Crawford Camiciottoli (eds.) (2015a), *Multimodal Analysis in Academic Settings: From Research to Teaching*, London: Routledge.

Fortanet Gómez, I. and B. Crawford Camiciottoli (2015b), "Introduction," in I. Fortanet Gómez and B. Crawford Camiciottoli (eds.), *Multimodal Analysis in Academic Settings: From Research to Teaching*, 1–14, London: Routledge.

Fricke, E. (2015), "Grammatik und Multimodalität," in C. Dürscheid and J. G. Schneider (eds.), *Handbuch Satz, Äußerung, Schema*, 48–76, Berlin: de Gruyter.

Goffman, E. (1983), "The Interaction Order," *American Sociological Review*, 48: 1–17.

Graßer, B. and A. Redder (2011), "Schüler auf dem Weg zum Erklären—Eine funktional-pragmatische Fallanalyse," in P. Hüttis-Graff and P. Wieler (eds.), *Übergänge zwischen Mündlichkeit und Schriftlichkeit im Vor- und Grundschulalter*, 57–78, Freiburg: Fillibach.

Greiffenhagen, C. (2014), "The Materiality of Mathematics: Presenting Mathematics at the Blackboard," *The British Journal of Sociology*, 65 (3): 501–28.

Gruber, H. (2012), "Funktionale Pragmatik und Systemisch Funktionale Linguistik—Ein Vergleich," in F. Januschek, A. Redder, and M. Reisigl (eds.), *Funktionale Pragmatik und Kritische Diskursanalyse*, 19–47, OBST 82.

Halliday, M. A. K. (1978), *Language as Social Semiotic. The Social Interpretation of Language and Meaning*, London: Edward Arnold.

Hanna, O. (2003), *Wissensvermittlung durch Sprache und Bild: Sprachliche Strukturen in der ingenieurwissenschaftlichen Hochschulkommunikation*, Frankfurt a.M.: Peter Lang.

Heller, D., A. Hornung, A. Redder, and W. Thielmann (2013), "The euroWiss-Project: Linguistic Profiling of European Academic Education (Germany/Italy)," *European Journal of Applied Linguistics*, 1(2): 317–20.

Iedema, R. (2001), "Resemiotization," *Semiotica*, 137(1/4): 23–39.

Januschek, F., A. Redder, and M. Reisigl (eds.) (2012), Funktionale Pragmatik und Kritische Diskursanalyse, OBST 82.

Kendon, A. (1980), "Gesticulation and Speech: Two Aspects of the Process of Utterance," in M. R. Key (ed.), *The Relationship of Verbal and Nonverbal Communication*, 207–27, De Haag: Mouton.

Krause, A. (2015a), "Medieneinsatz in Germanistik und Maschinenbau: Exemplarische Analysen von Vorlesungen," in M. Szurawitzki, I. Busch-Lauer, P. Rössler, and R. Krapp (eds.), *Wissenschaftssprache Deutsch: International, interdisziplinär, interkulturell*, 217–30, Tübingen: Narr.

Krause, A. (2015b), "Sprachliche Verfahren zur Vermittlung mathematischen Problemlösungswissens in der Hochschule—Exemplarische Analysen mathematischer Übungen," in G. Ferraresi and S. Liebner (eds.), *SprachBrückenBauen*, 203–18, Göttingen: Universitätsverlag.

Krause, A. and C. Carobbio (2014), "Sprachliche Verfahren der Vermittlung naturwissenschaftlichen Wissens in Physik-Vorlesungen: deutsch-italienische Perspektiven," *Studien zur deutschen Sprache und Literatur*, 32(2): 57–71.

Kress, G. (2014), "What is mode?" in C. Jewitt (ed.), *The Routledge Handbook of Multimodal Analysis*, 2nd edition, 60–75, London: Routledge.

Kress, G. and T. van Leeuwen (1996/98), *Reading Images. The Grammar of Visual Design*, London: Routledge.

Machin, D. (2014), "Multimodality and Theories of the Visual," in C. Jewitt (ed.), *The Routledge Handbook of Multimodal Analysis*, 2nd edition, 217–26, London: Routledge.

Maxwell, K. (2015), "When Here is Now and There is Then: Bridging the Gap in Time with "Sumer Is Icumen In,"" in J. Wildfeuer (ed.), *Building Bridges for Multimodal Research. International Perspectives on Theories and Practices of Multimodal Analysis*, 359–67, Frankfurt a.M.: Peter Lang.

Morris, C. W. (1938), *Foundations of the Theory of Signs*, Chicago: Chicago University Press.

Norris. S. (2004), *Analyzing Multimodal Interaction: A Methodological Framework*, London: Routledge.

Norris S. and C. D. Maier (eds.) (2014), *Interactions, Images and Texts. A Reader in Multimodality*, Berlin: de Gruyter.

O'Halloran, K. L. (2011), "Multimodal Discourse Analysis," in K. Hyland and B. Paltridge (eds.), *Continuum Companion to Discourse Analysis*, 120–37, London, New York: Continuum.

Redder, A. (1989), "Konjunktionen, Partikeln und Modalverben als Sequenzierungsmittel im Unterrichtsdiskurs," in E. Weigand and F. Hundsnurscher (eds.), *Dialonanalyse II*, 393–407, Tübingen: Niemeyer.

Redder, A. (2001), "Aufbau und Gestaltung von Transkriptionssystemen," in K. Brinker, G. Antos, W. Heinemann, and S. F. Sager (eds.), *Text- und Gesprächslinguistik/ Linguistics of Text and Conversation. Ein internationales Handbuch zeitgenössischer Forschung/An International Handbook of Contemporary Research. Halbband 2: Gesprächslinguistik*, 1038–59, Berlin: de Gruyter.

Redder, A. (2008), "Functional Pragmatics," in G. Antos and E. Ventola (eds.), *Handbook of Interpersonal Communication*, 133–78, Berlin: de Gruyter.

Redder, A. (2009), "Sprachliche Wissensbearbeitung in der Hochschulkommunikation – Empirische Analysen und kritische Perspektiven," in M. Lévy-Tödter and D. Meer (eds.), *Hochschulkommunikation in der Diskussion*, 17–44, Frankfurt: Peter Lang.

Redder, A. (2013), "Multilingual Communication in Hamburg – A Pragmatic Approach," in P. Siemund, I. Goglin, M. E. Schulz, and J. Davydova (eds.), *Multilingualism and Language Diversity in Urban Areas*, 259–87, Amsterdam: Benjamins.

Rehbein, J. (1984), "Remarks on the Empirical Analysis of Action and Speech. The Case of Question Sequences in Classroom Discourse," *Journal of Pragmatics*, 8(1984): 49 63.

Rehbein, J. (2001), "Konzepte der Diskursanalyse," in K. Brinker, G. Antos, W. Heinemann, and S. F. Sager (eds.), *Text- und Gesprächslinguistik/Linguistics of Text and Conversation. Ein internationales Handbuch zeitgenössischer Forschung/An International Handbook of Contemporary Research. Halbband 2: Gesprächslinguistik*, 927–45, Berlin: de Gruyter.

Rehbein, J. (2008), *Transcribing spoken language with EXMARaLDA/Partitur-Editor12—step by step instructions based on an example of spoken Turkish*, March 31. Available online: http://www.exmaralda.org/exmaralda/media/Transcribing-Spoken-Language-Turkish_EN.pdf.

Rehbein, J. and S. Kameyama (2004), "Pragmatik," in U. Ammon, N. Dittmar, K. Mattheier, and P. Trudgill (eds.), *Sociolinguistics/Soziolinguistik: An International Handbook of the Sciences of Language and Society/Ein internationales Handbuch zur Wissenschaft von Sprache und Gesellschaft*, 556–88, Berlin: de Gruyter.

Rehbein, J., T. Schmidt, B. Meyer, F. Watzke, and A. Herkenrath (2004), "Handbuch für das computergestützte Transkribieren nach HIAT," *Arbeiten zur Mehrsprachigkeit*, Folge B 56 1ff, March 31. Available online: http://www.exmaralda.org/files/azm_56.pdf.

Reisigl, M. (2012), "Epistemologische Grundlagen der Kritischen Diskursanalyse und Funktionalen Pragmatik," in F. Januschek, A. Redder, and M. Reisigl (eds.), *Funktionale Pragmatik und Kritische Diskursanalyse*, 49–71, OBST 82.

Scheflen, A. E. (1972), *Body Language and Social Order: Communication as Behavioral Control*, Englewood Cliffs: Prentice-Hall.

Schmitt, R. (ed.) (2007), *Koordination: Analysen zur multimodalen Interaktion*, Tübingen: Narr.

Schmitt, R. (2013), *Körperlich-räumliche Aspekte der Interaktion*, Tübingen: Narr.

Siefkes. M. and E. Arielli (2015), "An Experimental Approach to Multimodality: How Musical and Architectural Styles Interact in Aesthetic Perception," in J. Wildfeuer (ed.), *Building Bridges for Multimodal Research. International Perspectives on Theories and Practices of Multimodal Analysis*, 247–65, Frankfurt: Peter Lang.

Thielmann, W., A. Redder, and D. Heller (2015), "Linguistic Practice of Knowledge Mediation at German and Italian Universities," *European Journal of Applied Linguistics*, 3 (2): 231–53.

Tufte, E. (2003), *The Cognitive Style of PowerPoint*, Connecticut: Graphics Press.

van Leeuwen, T. (2005), *Introducing Social Semiotics*, London: Routledge.

Wildfeuer, J. (ed.) (2015), *Building Bridges for Multimodal Research: International Perspectives on Theories and Practices of Multimodal Analysis*, Frankfurt: Peter Lang.

7

Audio Description: A Practical Application of Multimodal Studies

Christopher Taylor

1 Introduction

Audio Description (AD) can be briefly defined as the insertion of short verbal descriptions illustrating the essential visual elements of an audiovisual product (including films, television programs, documentaries, and advertisements, but also such audiovisual phenomena as art galleries, museums, dance performances, city tours, and live events), principally for the visually impaired community, most succinctly described by Snyder (2008: 191) as "the visual made verbal." Clearly, in providing a verbal version of a visual text, the audio describer should ideally be versed in all the salient nonverbal characteristics of an audiovisual text. In the AD of films or other screen products, the dialogue remains unaltered, while the describer attempts to provide the sensorially disabled with a clear rendering of the visual components. Faced with numerous temporal and spatial constraints, the describer generally has to identify the most salient items for the description. This task would undoubtedly seem to require a certain knowledge of and a sensitivity toward what is now known as multimodal studies. Multimodality has been described as "the use of several semiotic modes in the design of a semiotic product or event" (Kress and van Leeuwen 2001: 20) and the use of "language, gesture, movement, visual images, sound and so on—in order to produce a text-specific meaning" (Thibault 2000: 311). It is not a new field of study in that everything is to some extent multimodal; but in the modern world, archetypal multimodal texts such as films, television programs, and websites have greatly broadened the scope of such studies. These media are also the main focus of audio description, and the first two in particular will be discussed in this chapter.

First the concept of audio description as a distinct text type will be addressed, as well as its role within the more established textual genre of film. This leads inexorably to a consideration of the essential question of what should be described by an audio describer. The issues involved are first illustrated with reference to the films *In the Name of the Father* (1993) and *Far from Heaven* (2002). Methods used to identify which elements to describe are then discussed, beginning with the use of eye-tracking technology and continuing with an illustration of the multimodal transcription technique. The limitations of the latter lead to a consideration of the less time-consuming technique of phasal analysis, which enables the describer to analyze a film with reference to such concepts as coherence, intertextuality, and visual and verbal cohesion. Examples of audio description are provided, ranging from the AD of the 2008 BBC TV series adaptation of the Dickens novel *Little Dorrit* to Stanley Kubrick's *The Shining* (1980), to the enigmatic movie *Memento* (2000) by Christopher Nolan. Finally, the European project Audio Description—lifelong access for the blind (ADLAB), designed precisely to investigate all aspects of audio description and to produce pan-European guidelines, is discussed in some detail. The "strategic" approach preferred by the producers of the ADLAB manual reflects all that transpires in the discussion of audio description presented in this chapter.

2 Audio Description as a text type

For a number of reasons, audio description represents a new language type, a genre in its own right. In terms of a "grammar" of audio description, unlike the much debated "grammar of film," which well transcends the linguistic sphere, certain specific grammatical characteristics can be noticed. There is an almost exclusive use of the present simple tense, with the occasional use of other present tenses (present continuous, present perfect). There is also an almost exclusive use of the third person, either actors (in the Hallidayan sense of animate or inanimate entities in various processes—see Halliday 1994) or pronouns. Clauses are exclusively declarative and there is a more than usual use of nonfinite clauses in theme position, the point of departure of the action, as exemplified by these clause openers from the film *The English Patient* (Minghella 1996):

"Seated behind the pilot, . . ."
"Forcing open a door, . . ."
"Wearing a simple skirt, . . ."

Film, too, is a broad text type, a complex semiotic mode making meaning in various ways, ways that need to be captured in audio description and fitted into the rather restricted genre structure described above. All films are to some extent dependent on an (undefined) number of structures, conventions, or frames, even to the point of cliché, and there is often a more or less explicit intertextuality present, both verbal and visual. The content of film dialogue, described as a "verbal audio channel" by Gottlieb (1998: 245), often refers, directly or indirectly, consciously or unconsciously, to other film texts or other sources. Sometimes this is self-contained in a series of films ("The name is Bond. James Bond") and other times a reference is made outside the film itself. In the film *Inglourious Basterds* (Tarantino 2009), Nazi officer Hans Landa greets his adversary with the words "Lieutenant Raines, I presume," clearly alluding to the famous greeting used by the explorer Stanley upon finding Dr. Livingstone in the African jungle.

Similarly elements of the "non-verbal audio channel" such as music can be intertextual, as in the use of Richard Strauss's *Also sprach Zarathustra* to introduce Stanley Kubrick's *2001: A Space Odyssey* (1968). "Verbal image" such as on-screen writing and "non verbal image," that is, all the iconic elements, are also inevitably prone to intertextuality to varying degrees. Place and time markers such as "Paris, 1789" and images such as the New York skyline are typical examples. Indeed only the most creative and innovative verbal or nonverbal creations completely escape the concept of intertextuality. The importance of these aspects of film text lies in whether they need to be mentioned (or not) by the audio describer, as they play an integral role in the aforementioned "frame" and may be important to a fuller understanding of the text. There are essentially two schools of thought in this regard, which can be roughly described as the American and the European. The former tends to eschew any kind of interpretation on the part of the describer and any form of "appraisal" (see Martin and White 2005), maintaining that "what you see is what you say" (WYSISWYS) is the approach to follow. Many European exponents and scholars tend toward more flexibility in this regard and may add elements of appraisal ("a pretty girl," "a menacing look") and also explain any example of intertextuality that may not be immediately understandable to the (visually challenged) audience, either because of lack of world knowledge or perhaps a lack of sensitivity to cultural references. But there is indeed a fine line between judicious intervention and unwanted interpretation. This remains a controversial issue among different AD schools and will be addressed again later in this chapter.

2.1 Audio Description: What to describe

The question of what to describe in an AD is of course the fundamental point of interest. Where should the focus lie among the myriad semiotic modalities present and operating in a multimodal text? Which are the salient images, sounds, or symbols? Clearly in any frame, shot, or scene in a film, there are too many elements present to be perceived in totum by any viewer and most of the "redundant" items do not need describing.

2.1.1 *In the Name of the Father*

A film that has proved very useful in explaining the resources of multimodality is *In the Name of the Father* (Sheridan 1993). The opening sequence brings together a wealth of semiotic modalities, making a powerful impact on the audience. It also follows a filmic style seen in a considerable number of other films. Within minutes of the beginning of the film, there is a violent explosion, in this case the Irish Republican Army bombing of a pub in Guildford (cf. the beginning of *Selma* (DuVernay 2015), where a church is blown up, and the beginning of *Die Hard with a Vengeance* (McTiernan 1995) in which a bank is blown up). Bateman (2015), with reference to the latter film, discusses regularity of form and Hollywood continuity and how certain filmic structures occur and reoccur in predictable patterns.

To return to *In the Name of the Father*, the opening scenes are set in strife-torn Belfast during the "Troubles," and contain dialogue, contrasting northern Irish accents with the London-area voices of the soldiers, music in the form of an introductory ballad sung by (the Irish) Bono, and the heavy rock music of Jimi Hendrix circa 1974, which provides the perfect accompaniment to the street chase scene. Written language features in the caption "Guildford, England 1974" and in the graffiti on the walls of Belfast "BRITS OUT!" Several of the characters reveal their Irishness in the color of their hair and their state of poverty by the clothes they wear. Symbols are used—a wig to denote a lawyer, an ice-cream van to denote the period—and facial expressions and body language reveal a range of emotions from fear to excitement to rage. All of this is immediately perceived by a normally sighted audience, but the way to describe these features to the visually challenged requires an understanding of multimodality and the integration of semantic modes that transcends the canonical "Say what you See."

2.1.2 *Far from Heaven*

Similarly, the opening sequence from the film *Far from Heaven* (Haynes 2012) contains symbols that help establish era, season, place, mood, etc. The initial street scene featuring classic makes of car, men in trilbies, and ladies in flouncy dresses all point to the 1950s in small town America. The film title and the names of the actors appear in the kind of colorful, birthday card-type letters that were typical of that era. The yellow or brown leaves tell us it is autumn, and everything seems to be in its place in the peaceful scene of small town prosperity with which the audience is presented. The creating of this atmosphere is important in that it will be shattered by events to come. But what is the most effective way of describing such features as those mentioned above in order to give the visually challenged audience the same experience as the normally sighted public?

2.2 Eye tracking

Identifying what is redundant is a major step in the analysis of a text to be described. Several approaches have been adopted in an attempt to achieve this end. One of the most promising has been the adaptation of eye-tracking technology and technique to audio description. The practice of eye tracking has been used successfully in various fields, for example in analyzing perception in subtitled films (d'Ydewalle and de Bruycker 2007; Perego 2012; Schotter and Rayner 2012), also with an eye to second language learning (d'Ydewalle and Pavakanun 1997; Roberts and Siyanova-Chanturia 2013). It has subsequently been harnessed to AD studies as a means of identifying the seemingly most salient visible elements in a video, at least as regards people with normal vision. Fixations and saccades (the gaps between fixations) track the eyes' route up and down and around the image; the fixations indicate points of focus and the saccades indicate what appears to arouse most interest.

Figure 7.1 is a still showing two Japanese women at work and the major and minor eye movements and fixations of a viewer. It would seem that the faces of the women are important, especially the movement of their hands. The eyes move back and forth between these elements, and much of the rest of the image is virtually ignored. In Figure 7.2 the "heat spots" show the concentration of the viewer on said elements.

The importance of movement is also revealed in the eye tracking of a scene from the film *Marie Antoinette* (Coppola 2006), where concentration on the Queen's kicking feet (Figure 7.3) led to viewers not noticing the anomalous (and erroneous) presence on screen of a pair of modern sneakers in this eighteenth-century scenario. These studies at least tell us how normally sighted viewers "see" a film and that movement should be a priority in an audio description.

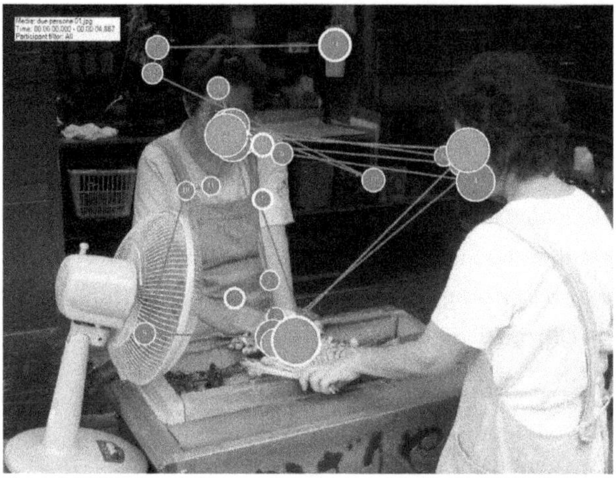

Figure 7.1 Eye tracking of Japanese women.

Figure 7.2 Eye tracking—heat spots.

2.3 Multimodal metonymy

Another interesting approach to examples such as Marie Antoinette's feet is to be found in the work of Forceville (2009) and Moya Guijarro (2015), who both discuss the concept of multimodal metonymy. Guijarro cites a picture from the Beatrix Potter children's book *Peter Rabbit* in which we see only the boot of an enraged farmer who is attempting to kick Peter out of his shed. Guijarro argues that the metonymic portrayal of the boot, which represents as a part-whole relationship both the farmer and his frame of mind in that moment, is a more

Figure 7.3 Eye tracking a scene from *Marie Antoinette.*

powerful image than a picture of the farmer in his entirety. Marie Antoinette's feet, twitching in childish glee at yet another pair of fancy shoes, provides us with an image that portrays more than just the Queen's appendages: it captures her frivolous approach to unlimited luxury and total disinterest in weightier matters. If this is not clear from the "say what you see" approach, in these cases the visually challenged audience might also need some assistance.

In another approach, Bateman and Tseng (2015: 139), when analyzing the opening shots of the Hitchcock film *The Birds* (1963), point out that during the initial street scene involving Melanie, the famous character played by Tippi Hedren, "individual cars, other passers-by, fire hydrants, etc." are not crucial to the cohesive structure of the film text. In this case it is because of the fact that "they do not reappear [so] that they will not participate in cohesive ties" (Bateman and Tseng 2015: 139). Such elements can, therefore, often be considered expendable in an audio description.

3 A case study

There are, however, many other factors involved. For example what lies behind the immediate visuals? A good example of a successful audio description is that of the BBC's serialized version of the Charles Dickens novel *Little Dorrit*. The novel started life as a magazine serial between 1855 and 1857 before being published in a single volume in 1857. The BBC serial was produced in 2008 and audio description was added to the DVD version issued in 2009. Andrew Davies

(2008), who adapted the story for television, claimed that one particular image was dominant in his mind, and this was of "Little Dorrit going out in the early morning, emerging from the gates of the Marshalsea prison, where she lived with her incarcerated father, hurrying through the mean streets, with the dark, gloomy buildings looming over her" (Davies 2008: xvii). In the novel Dickens describes this reality:

> An oblong pile of barrack building, partitioned into squalid houses standing back to back . . . hemmed in by the high walls duly spiked at top. (Dickens 2008 [1857]: 57)
>
> She had begun to work beyond the walls . . . to come and go as secretly as she could between the free city and the iron gates. Her original timidity had grown . . . and her light step and her little figure shunned the thronged streets while they passed along them.
>
> This was the life of Little Dorrit, turning at the end of London Bridge . . . (Dickens 2008 [1857]: 78)

The first scene in the DVD, one which precedes the opening credits, is of the birth of Amy (Little) Dorrit in the prison. The audio description begins with a time and place marker: "1808 Marshalsea Debtors Prison." This is followed by a description of the action taking place on screen: "The doctor shows the baby to a boy and girl." Then the credits are shown and audio-described: "Old discs spinning on lengths of thread in the darkness – (list of names) BBC Little Dorrit by Charles Dickens."

This particular line has been criticized (by some normally sighted viewers) as being too decontextualized and difficult to perceive. However, the whole AD has been well received by visually challenged audiences. Many in the visually challenged audience are used to this kind of introduction to a film and can contextualize such text with comparative ease. A time marker then appears on the screen as a caption and is dutifully read for the audience: "21 years later." After this time lapse, the AD moves into character description: "A neatly dressed man in his 20s with short brown hair unlocks a small wooden door." And then Amy: "Wearing a smart grey dress and white pinafore, a straw bonnet and light blue cape, Amy climbs through the door carrying a wicker basket and walks into the busy streets outside"

The description that follows attempts to create the image that Davies had in mind: "She passes a horse-drawn carriage and two men in top hats on their way across a bridge." The bridge in question is the London Bridge mentioned in the novel. However, on this occasion the describer does not limit himself to mere description but provides a beautifully succinct appraisal of Little Dorrit's life and role in the big city. "A tiny figure dwarfed by its enormity. . . ." Such liberty of expression would be frowned upon by some exponents of AD, particularly those of the American school who support the slogan "What you Say is what you See" (Snyder 2008: 197). And yet, this kind of approach can answer the question posed above, "What lies behind the immediate visuals?"

Bateman (2013: 139), following (and supporting) Metz's early semiotic film theory (1974), talks of "functional units of discourse" as a way of finding a parallel to the semantic units of text described in the Hallidayan tradition (see Halliday 1978). However these units are to be found at a more abstract level than the classic division of film into shots and scenes. The unit of filmic "discourse" may overlap a succession of shots or even scenes. For this reason, it may be advisable for the audio describer, at certain points during the unraveling of the film narrative, to provide more information than simply "what you see." The filmmaker is "making meaning" with his or her product in an attempt to explain, convince, persuade, entertain a potentially vast and varied public. The functional units originate in the filmmaker's mind and, delving into the terminology of systemic functional semiotics, are "realized" or "instantiated" as the material units of film (frames, shots, sequences). Although there is a strong relationship between these two strata, which when woven together create the semiotic texture of the film product, the syntagmatic progression or structure and the paradigmatic choices made may not exactly reflect the more abstract unit of film discourse as conceived by the filmmaker.

In the excerpt from Little Dorrit referred to above, the first part of the first shot showed a neatly dressed man in his twenties with short brown hair unlocking a small wooden door, which was audio-described precisely as "A neatly dressed man in his 20s with short brown hair unlocks a small wooden door."

First, the description was selective—it made no mention of the actual clothes worn by the man, their color, his height, etc. and no mention of any surrounding features. This was presumably the result of the describer's appraisal of the salient items to describe, based on intuition, experience, or even resort to the literature now available on AD strategies. Second, the description reflects the actual filmic material visible on the screen—it has gone from the verbal (it was originally scripted in some way) to the visual, back to the verbal as a sentence. The action continues: the young man greets Amy Dorrit as she comes out of the "small wooden door" to venture forth into the streets of London. No audio description is heard as the dialogue needs to be heard and indeed explains the activity: "Good morning, Amy"—"Good morning, John." And the film goes on. However, at a higher level of abstraction, as intended originally by Dickens and here portrayed by Davies, it is clear from the body language that John is attracted to Amy. This is to be an important element in the story and needs to be established from the beginning. It is, thus, arguable that the visually challenged audience needs some assistance in deciphering this fact. It is also true that they may glean something of this nature from John's tone of voice with his "Good morning, Amy," but the question is still worth raising. The multimodal text holds many secrets from an audience that is denied access to part of that multimodality, and the describer needs to consider all options.

3.1 Multimodal transcription

A further consideration when discussing film analysis is whether it is possible to identify reoccurring structures in the same way as verbal language structures reoccur. Individual film frames can have infinite variation whereas language is more circumscribed by rules of grammar and syntax. And yet repeated scene types, film introductions, dialogue sequences, and so on can be recognized. As an example, Stanley Kubrick's film *The Shining* (1980) contains a classic threatening confrontation scene featuring an insanely violent Jack Nicholson (see Figure 7.4). Because this type of situation (the row) has appeared in innumerable films, it can be argued that it is easily "visualizable" by a visually challenged audience. And this may well be the case in many such examples, and the audio describer can be economic with his or her description, especially as the dialogue is often dense and rapid. In the case of the scene from *The Shining*, however, the visuals are accompanied by gestures, facial expressions, sounds, and eerie music, which may need explaining to a visually challenged audience.

Figure 7.4 Jack Nicholson in *The Shining*.

The shouting and threatening language that can be heard may not be sufficient to gauge the level of irrational madness.

Another approach to identifying the salient elements to describe in an AD is through a multimodal transcription. The multimodal transcription, as devised by Thibault (2000) for his analysis of the multimodal structure of a bank advertisement and adapted by Taylor (2009) to provide a rationale for choices made in subtitling film, can prove useful also for the audio describer.

Table 7.1 shows a multimodal transcription of a small part of the row scene in *The Shining*. For a thorough explanation of the coding system and the modalities used in the analysis, see Thibault (2000). Suffice it, for this account, to point out that a minute description of what is happening on screen can help in the process of condensing subtitles where the time factor demands it. If meaning is conveyed through gesture, body language, on-screen clues, and so on, the subtitles of the dialogue can be appropriately shortened. As regards audio description, on the other hand, the frame-by-frame multimodal transcription distinguishes between the salient items and the secondary features, thereby indicating the most important images to describe.

3.2 Phasal analysis

However, it should be apparent that the multimodal transcription is extremely time-consuming and impractical for texts other than very short video products, though very useful for sensitizing students to the mechanisms of multimodal texts. Thus, in an attempt to find a more amenable analytic approach, an extension of the rationale that lies behind the multimodal transcription can be found in the concept of phasal analysis. This kind of analysis is illustrated in Gregory (2002) in relation to written language, where he explains that "phase characterizes stretches within discourse (which may be discontinuously realized in text) exhibiting their own significant and distinctive consistency and congruity in the selections that have been made from the language's codal resources" (Gregory 2002: 321).

Reducing the concept to the sole identification of discrete phases (including the idea of macrophases and subphases), multimodal texts can be described in this phasal fashion. Thibault talks of television advertisements in which "there are typical, phase-specific ways in which, say, speech, chorus and music interact both with each other and with particular kinds of action sequences and participant categories in the visual semiotic[, and] there are readily identifiable typical combinations of multimodal genres, e.g. logo, graphological resources of written language, musical closure" (Thibault 2000: 366).

Table 7.1 Multimodal transcription from *The Shining*

T	Visual frame	Visual image	Kinesic action	Sound track	Subtitles
	Shot 1				
		CP: moving forward HP: frontal VC: Wendy's dress reminding us of a squaw's. Entrance to the lounge and stairs in the background. Sunlight coming in through big windows. Carpets reminding us of the Indian culture. VS: Wendy. Baseball bat (protection). The flight of stairs. CO: naturalistic CR: light green and beige VF: Close, other participant (eye contact)	Walking backward, she grimaces with fear, shaking her head, holding the bat (slightly waving because of her walking). Tempo: S	(RG) (p) (Wendy) Jack Please Volume: n Tempo: M (PAUSE)	Jack, per favore
	Shot 2				
		CP: moving backward HP: slightly oblique VC: Jack's work clothes (Western kind of clothes). Sunlight coming in. Background slightly out of focus. VS: Jack's reaction CO: naturalistic CR: bright and dark colors VF: Close, other participant (eye contact)	Walking forward, Jack bends slightly, head thrust forward and slightly to the left, eyes challenging Wendy (slightly frowning)	(RG) (p) (Jack) You believe his	Pensi che potrebbe

Shot 3				
	CP: moving backward HP: slightly oblique VC: Jack's work clothes (Western kind of clothes). Sunlight coming in. Background slightly out of focus. VS: Jack's reaction CO: naturalistic CR: bright and dark colors VF: Close, other participant (eye contact)	(as above and Jack starts grinning)	health might be	ammalarsi

This approach enables the analyst to identify homogeneous "phases," both continuous and discontinuous, within a multimodal text and to identify the distribution of the semantic components of film discourse and recognize register changes, character traits, and elements of cohesion and coherence. While the multimodal transcription reveals these resources in detail frame by frame, phasal analysis allows us to determine the basic constituent features and organization of the phases that structure the information content of the film. These phases, which can be divided into macrophases, phases, subphases, and so on, are identified by examining the portions of text containing semiotic resources with a high level of cohesion in terms of both vision and sound, elsewhere described as "covariate semantic ties" (Lemke 1985: 287). In other words, a scene featuring a particular set of characters discussing a particular issue in a particular setting accompanied by a particular piece of music can constitute a phase. Such a portion of a film text can be seen to be characterized by a high level of coherence and metafunctional homogeneity. That is, in the social semiotic of systemic-functional parlance, cohesive ideational, interpersonal, and textual components of single stretches of text can be labeled as phases. These phases are often discontinuous in the sense that a scene involving two bank robbers escaping from a robbery may be revisited several times while interspersed by other, also discontinuous, phases involving police officers conferring, cars chasing through city streets, etc.

By way of example, in the film *Memento* (Nolan 2000), it is initially possible to identify two macrophases in that the first part of the film is shot in black and white and follows events in chronological order, while the second part is in color and presents events in retrospect. Macrophase 1 is then subdivided into three phases, and Macrophase 2 into eleven such phases. Returning to Macrophase 1, the phases are further divided in subphases as shown in Table 7.2. Many of the phases are discontinuous in nature in that the various episodes of which they are composed alternate throughout the film before coming together at the end.

An even more delicate analysis can be made, but this phasal breakdown is sufficient to understand the narrative structure and meaning of this rather obscure film. To simplify the analysis in this case (see Vidmar 2005), and so as to make an immediate distinction between the various phases, each phase is defined on the basis of its ideational content (a place, an object, a person, an action, or a narrative strategy such as black-and-white footage or flashback). It is important to stress that the components of all the various phases are examined, in that it is necessary to not only trace the distinct elements that determine internal homogeneity in a specific phase but also to identify the more general characteristics that show what each subphase has in common with other

Table 7.2 Phases in *Memento*

Macrophase 1	Phases	Subphases
Black and white	P1: Leonard in the hotel room	SP1: Leonard thinking
		SP2: The telephone rings and Leonard responds
		SP3: Leonard speaks on the phone
		SP4: The telephone rings and Leonard doesn't respond
		SP5: The caller hangs up
		SP6: Tattoo on the leg
		SP7: The hotel manager
		SP8: An envelope pushed through the door
	P2: Leonard outside the hotel room	SP1: Leonard and Teddy
		SP2: Leonard and Jimmy
	P3: Flashback	SP1: Leonard's past
		SP2: Sammy Jankis

subphases, thereby indicating the continuity and cohesion of the information as it unfolds in the film.

Starting with Macrophase 1, *Black and White*, one of the two temporal sequences in which the film is structured, the character Leonard is introduced. The distinct elements contained in this Macrophase are the black-and-white images, the chronological sequencing of events, and the fading in and out of the scenes. In Phase 1, *Leonard in the hotel room*, the things that happen to Leonard in the hotel room at the Discount Inn are described. The distinct elements are the room, Leonard's clothing (a checked flannel shirt and white boxer shorts), and an undefined noise that may be coming from the air-conditioning system. In Subphase 1, *Leonard thinking*, Leonard is in the room but does not know where he is or why he is there. He searches the room to find an answer. The audience hears what Leonard is thinking. The distinct element is Leonard's off-screen voice. And so on.

This phasal description enables us to observe the hierarchical organization of the narrative content of the film and how the myriad semiotic resources connect within the dynamic progress of the communication. For the audio describer, phasal analysis provides a further tool for the identification of those salient elements among the many visual components, and also some nonvisual components, by stripping the film text down to its constituent parts and examining how those parts intertwine in continuous or discontinuous phases.

4 ADLAB

The importance of audio description in the general debate surrounding the question of access, particularly access to audiovisual products on the part of the sensorially disabled, led a number of scholars, who had been involved in audiovisual studies for some time, and three service providers to embark on a European-funded project designed to investigate more thoroughly all the issues discussed so far in this chapter. Thus the idea for ADLAB became, from 2011 to 2014, a European Lifelong Learning project coordinated by the University of Trieste in Italy. The project studied AD in depth and produced a book of theory (Maszerowska, Matamala, and Orero 2014) and a set of strategic guidelines for the profession (Remael, Reviers, and Vercauteren 2015). This research investigated the myriad of ways in which multimodality impacts the analysis of texts such as film and television products. As explained previously, audio description involves the study of all the semiotic resources in a text and how they interact, to create a vehicle that, in addition to providing an essential tool in promoting the social inclusion of the visually challenged community, makes a contribution to harnessing our understanding of how multimodality influences the analysis of texts such as film and television products. In the process of identifying best practices in the creation of audio descriptions, it was decided to use a single film vehicle for the various stages of analysis, testing, and describing. The choice fell on *Inglourious Basterds* (Tarantino 2009) because of the extreme difficulty it poses for a host of reasons. The project report refers to the combination of languages spoken by the characters, and the direct implication for the translation and audio description was perhaps the first reason to suggest this title. The plot at some turning points develops from the language characters speak, and by the cultural references or gestural features within the language. These instances were important to test the different audiovisual translation modes and their direct implication for audio description possibilities and strategies.

4.1 *Inglourious Basterds*

Inglourious Basterds was also chosen because it encompasses many film genres, from Westerns to James Bond movies, and elements of intertextuality therein, as described earlier, are manifold. At the very beginning, the theme music is the same as that for the classic 1960 John Wayne movie *The Alamo*, cementing the Western connection. This is reinforced shortly after when music in the style of

Ennio Morricone is featured. In between these Western sounds, the beautiful strains of Beethoven's *Für Elise* are heard, providing a horrifically ironic accompaniment to the threatening behavior of a ruthless Nazi officer. Direct or indirect reference is also made to other films, film directors, and film characters such as Leni Riefenstahl and Michael Myers. These intertextual allusions and references were important to the film and represented one of the questions that had to be answered on each occasion regarding the amount of information to give a visually challenged audience without sliding into condescension. Similarly, in terms of subjective appraisal, while at times it was important to portray what was seen and heard on the screen, at other moments the focus had to be placed on the emotions provoked. The film recounts an emotionally and historically potent story where clear emotions are present in the film characters and these emotions are transmitted to the audience. Here audio describers have to tread a fine line between objective and subjective interpretations. The visual, audio, and narrative portrayal of violence, love, hate, disgust, or horror needs to be described, and the describer will need to prioritize and disambiguate information, without pandering to the visually challenged audience. But, for example, when a German soldier's skull is crushed by a baseball bat, is that horrendous sound perceivable by the non-seeing public, or does it need to be explained?

The film was also chosen because of the particular editing techniques it employs. Shot changes are frequent and often move deftly in time and space. The camera movements are particularly effective in illustrating facial expressions and gestures. The crucial scene in the restaurant basement hinges on a fatally mistaken gesture. Both color and black-and-white scenes are used, especially in the "film within the film" sequences.

And so the challenge was how to capture the many layers of information and to discriminate between them in order to create a cohesive and meaningful audio description. The individual members of the project worked on creating a matrix of all the critical points encountered in writing an audio description of this film, and this led to the structuring of a volume in chapters dealing with distinct aspects. From the description of the relevant images, the discussion moved to dealing with written text appearing on screen, cultural references, spatiotemporal features, and the importance or otherwise of secondary elements. The question of how to describe certain sounds that would not be immediately apparent to a visually challenged audience and the problems involved in fixating and describing all the characters in a film were tackled. Gestures and facial expressions were afforded a separate chapter, as was the use of the metalanguage of film. The previously mentioned phenomenon of intertextuality was discussed along with elements of textual or visual cohesion. Finally, Audio Introductions was included as a useful adjunct to the AD process, and the necessity for Audio Subtitles was realized as it received space as a useful element in the describing of foreign films.

4.2 The manual

The final stage in the process was the writing of the manual, coordinated by colleagues at the University of Antwerp and consisting of approximately a hundred pages of "advice," in the sense that prescriptive rules were bypassed (they already exist) and "strategic guidelines" proposed instead. By way of example, the question of how to describe characters cannot be handled by fixed rules as a number of factors are involved in the decision-making process. There is always a paradigmatic choice to be made between alternatives when, for instance, introducing a character for the first time. In the case of a known historical or contemporary figure, this can depend on whether the audience can be expected to know the character or not but also on the space available or on the general world knowledge of the audience. In the film *The Hours* (Daldry 2002), the main character is the doomed writer Virginia Woolf, played by Nicole Kidman. The immediate options available to the describer when Virginia first appears at the beginning of the film are:

- simply name the character: Virginia Woolf
- provide a gloss and name the character: the English writer Virginia Woolf
- describe the character in detail without naming her: a middle-aged woman with a slightly hooked nose and hair pulled back in a bun
- describe the character in detail and name her: a middle-aged woman with a slightly hooked nose and hair pulled back in a bun, Virginia Wolf

In the first case the assumption made is that the audience knows that the film is about Virginia Woolf and the audience, or at least most of the potentially large audience, know who she was. Such a choice could also be occasioned by lack of time. The second option would be used as a further guarantee of audience recognition of the character, and if time permits it. The third choice, which requires sufficient time, would be used if the name of the character were not considered important to the storyline of the film and would not be used, for example, in *The Hours*. The fourth option is an indulgence that can be used if the time available is sufficient and the naming, in apposition, is used for effect. There could be other versions, and the manual encourages the describer to consider all options before making a choice.

Returning to the question of verbal or visual intertextuality, the allusion to other texts (books, films, sayings, etc.) is exemplified by a scene from the film *Hitchcock* (Gervasi 2012). At the end of the film, which has focused on the story behind Hitchcock's highly successful *Psycho*, the famous director says, "Unfortunately, I find myself once again bereft of all inspiration. I do hope something comes along soon." At this point a bird lands on his arm. The important question here for the audio describer is to decide how much help the audience requires in perceiving the connection to his next film *The Birds* (1963). Prescriptive rules such as "Do not refer to intertextual ties," "Make intertextual ties explicit," and "Omit intertextual references" are not helpful. Thus, the ADLAB guidelines would encourage a kind of internal conversation:

- Use a simple description if you can assume the audience sees the connection with Hitchcock's *The Birds*, such as, "A bird lands on his arm."
- If you are not certain, but you think the audience will realize the connection, you can use an enhanced description such as, "A bird lands on his arm, as a sign of things to come."
- If you are certain that the audience will not make the connection, and you consider this scene important, then you can make it explicit, such as, "A bird lands on his arm, alluding to his next film, *The Birds*" (AD linking marker and marked).
- NB: Your knowledge of the particular situation (who is likely to make up most of the audience, the likelihood that visually challenged patrons are familiar with 1960s films, etc.) will affect your judgment here rather than hard-and-fast rules.

5 Conclusion

The practice of audio description, as described in the opening section of this chapter, brings the describer face to face with the concept of multimodality. The decisions the describer has to make when providing a verbal exposition of the images that constitute a film or other screen product involve a judicious selection of semiotic resources. Even nonvisual aspects such as unclear noises or intertextual links may need describing, together with the images and on-screen written material, and all within the time constraints inherent to this practice. Various tools are available to the audio describer to assist him or her in making appropriate decisions and identifying the most salient features. These include eye-tracking technology, which traces the focus of normally sighted individuals; the multimodal transcription, which minutely examines the content of a film, frame by frame; and phasal analysis, which enables the describer to understand the structure of film and to identify the most meaningful elements. But in the final analysis it is the describer who must make decisions based on the particular circumstances surrounding the task in hand. For this reason the European project ADLAB made its ultimate objective the production of a detailed manual providing the describer with strategic guidelines designed to get him or her to consider alternative solutions depending on the specific circumstances of the case. These could be an appraisal of the local or world knowledge of the visually challenged audience in question, a consideration of the need to provide extra information in order to clarify certain filmic episodes, risking decisions that might lead to the accusation of condescension toward the audience, and so on. Having eschewed the idea of hard-and-fast prescriptive rules, the ADLAB guidelines often appear as pieces of advice on how to handle particular situations—situations that occur and reoccur in films of all genres. The research conducted during the project lifetime, which prior to the issuing of the manual had led to the publishing of a book on the many theoretical and practical aspects of audio description, investigated the myriad ways in which multimodality impacts on the analysis of texts such as film and television products. This chapter has attempted to show that audio description involves the study of all semiotic resources in a text and how they interact, providing a vehicle that, in addition to providing an essential tool in promoting the social inclusion of the visually challenged community, also makes a contribution to harnessing our understanding of multimodality.

References

Bateman, J. (2013), "Hallidayan Systemic-Functional Semiotics and the Analysis of the Moving Audiovisual Image," *Text and Talk*, 33: 641–63.

Bateman, J. (September, 2015), Defining mode – but this time for real! Paper presented at the Second Bremen Conference on Multimodality. Bremen.

Bateman, J. and C. Tseng (2015), "The Establishment of Interpretative Expectations in Film," in M. J. Pinar Sanz (ed.), *Multimodality and Cognitive Linguistics*, 131–46, Amsterdam: John Benjamins.

Davies, A. (2008), "Introduction," in C. Dickens (2008 [1857]), *Little Dorrit*, xv–xviii, London: Ebury Publishing.

Dickens, C. (2008 [1857]), *Little Dorrit*, London: Ebury Publishing.

d'Ydewalle, G. and U. Pavakanun (1997), "Could Enjoying a Movie Lead to Language Acquisition?" in P. Winterhoff-Spurk and T. H. A. van der Voort (eds.), *New Horizons in Media Psychology*, 145–55, Opladen: Westdeutscher Verlag.

d'Ydewalle, G. and W. de Bruycker (2007), "Eye Movements of Children and Adults while Reading Television Subtitles," *European Psychologist*, 12 (3): 196–205.

Forceville, C. (2009), "Metonymy in Visual and Audiovisual Discourse," in E. Ventola and A. J. M. Guijarro (eds.), *The World Told and the World Shown. Issues in Multisemiotics*, 56–74, Basingstoke/New York: Edward Arnold.

Gottlieb, H. (1998), "Subtitling," in M. Baker (ed.), *Routledge Encyclopedia of Translation Studies*, 244–48, London, New York: Routledge.

Gregory, M. (2002), "Phasal Analysis within Communication Linguistics," in P. H. Fries, M. Cummings, D. Lockwood and W. Spruiell (eds.), *Relations and Functions Within and Around Language*, 316–45, London: Continuum.

Halliday, M. A. K. (1978), *Language as Social Semiotic*, London: Edward Arnold.

Halliday, M. A. K. (1994), *An Introduction to Functional Grammar*, London: Edward Arnold.

Kress, G. and T. van Leeuwen (2001), *Multimodal Discourse: The Modes and Media of Contemporary Communication*, London: Edward Arnold.

Lemke, J. (1985), "Ideology, Intertextuality and the Notion of Register," in J. Benson and W. Greaves (eds.), *Systemic Perspectives on Discourse, Volume 1. Selected Theoretical Papers from the 9th International Systemic Workshop*, 275–94, Norwood, NJ: Ablex.

Martin J. R. and P. White (2005), *The Language of Evaluation: Appraisal in English*, Basingstoke: Palgrave Macmillan.

Maszerowska, A., A. Matamala, and P. Orero (eds.) (2014), *Audio Description: New Perspectives Illustrated*, Amsterdam/Philadelphia: John Benjamins.

Metz, C. (1974), *Film Language: A Semiotics of the Cinema*, Oxford: Oxford University Press.

Moya Guijarro, J. (2015), "Visual Metonymy in Children's Picture Books," in M. J. Pinar Sanz (ed.), *Multimodality and Cognitive Linguistics*, 115–30, Amsterdam: John Benjamins.

Perego, E. (ed.). (2012), *Eye-Tracking in Audiovisual Translation*, Rome: Aracne Editrice,

Remael, A., N. Reviers, and G. Vercauteren (eds.) (2015), *Pictures Painted in Words: ADLAB Audio Description Guidelines*, Trieste: EUT.

Roberts, L. and A. Siyanova-Chanturia (2013), "Using Eye-Tracking to Investigate Topics in L2 Acquisition and L2 Processing," *Studies in Second Language Acquisition*, 35(2): 213–35.

Schotter, E. R. and K. Rayner (2012), "Eye Movements in Reading. Implications for Reading Subtitles," in E. Perego (ed.), *Eye-Tracking in Audiovisual Translation*, 83–104, Rome: Aracne Editrice.

Snyder, J. (2008), "The Visual Made Verbal," in J. Diaz-Cintas (ed.), *The Didactics of Audiovisual Translation*, 191–98, Amsterdam: John Benjamins.
Taylor, C. (2009), "Pedagogical Tools for the Training of Subtitlers" in J. Díaz Cintas and G. Anderman (eds.), *Audiovisual Translation: Language Transfer on the Screen*, 214–28, Basingstoke: Palgrave Macmillan.
Thibault, P. (2000), "The Multimodal Transcription of a Television Advertisement: Theory and Practice," in A. Baldry (ed.), *Multimodality and Multimediality in the Distance Learning Age*, 311–85, Campobasso: Palladino Editore.
Vidmar, V. (2005), *Analisi fasale e multimodale di un film: Memento*, PhD diss., University of Trieste, Trieste.

Filmography

2001: A Space Odyssey (1968), [Film] Dir. S. Kubrick, USA: MGM.
Die Hard with a Vengeance (1995), [Film] Dir. J. Tierman, USA: Fox.
Far from Heaven (2002), [Film] Dir. T. Haynes, USA: Focus Features.
Hitchcock (2012), [Film] Dir. S. Gervasi, USA: Fox Searchlight Pictures.
Inglourious Basterds (2009), [Film] Dir. Q. Tarantino, USA: Alliance Films.
In the Name of the Father (1993), [Film] Dir. J. Sheridan, USA: Universal Pictures.
Marie Antoinette (2006), [Film] Dir. S. Coppola, USA: Sony/Columbia.
Memento (2000), [Film] Dir. C. Nolan, USA: Newmarket Films.
Psycho (1960), [Film] Dir. A. Hitchcock, USA: Paramount Pictures.
Selma (2015), [Film] Dir. A. DuVernay, USA: Paramount Pictures.
The Alamo (1960). [Film] Dir. J. Wayne, USA: The Alamo Company.
The Birds (1963), [Film] Dir. A. Hitchcock, USA: Universal.
The English Patient (1996), [Film] Dir. A. Minghella. USA: Miramax Films.
The Hours (2002), [Film] Dir. S. Daldry. USA: Paramount Pictures.
The Shining (1980), [Film] Dir. S. Kubrick. USA: Warner Bros.

8

Multimodal Translational Research: Teaching Visual Texts

Victor Lim-Fei and Serene Tan Kok Yin

1 Introduction

Our world is changing every day with the affordances brought about by digital media and technology. With the ubiquity of multimodal texts and the different ways of meaning making, the reality of multimodal communication has grown increasingly apparent. The traditional focus on literacy and numeracy will no longer be sufficient for students to navigate the complex multimodal communicational landscape that they inhabit. New skills for reading and finding, authenticating, linking, and representing information are demanded in this increasingly interactive digital media-enabled multimodal environment (Kress 2003; Jewitt 2007).

This chapter describes the translational research efforts made to apply the theories and frameworks developed in academia to inform teaching and learning in the secondary education setting. This follows the direction of "appliable discourse analysis" (Matthiessen 2013), where, in this case, multimodal discourse analysis frameworks and approaches are translated into instructional strategies for the classroom. By drawing on the multimodal discourse frameworks developed by O'Toole (2010), Kress and van Leeuwen (2006), and later extended by Lim-Fei and O'Halloran (2012), Tan, Marissa and O'Halloran (2012), O'Halloran and Lim-Fei (2014), and Lim-Fei et al. (2015), the authors have developed an approach to the teaching of visual texts aimed at secondary school students in Singapore. Visual texts are defined as discourses that are constructed using only images or that have a combination of image(s) and written or oral language. Examples include advertisements and posters—print and electronic. This chapter describes the translational process from theories in the field of multimodality to a set of instructional strategies for the teaching of visual texts.

2 Multimodal literacy

Information, particularly in the digital age, is represented not just with language alone. Instead, language is often nestled among other semiotic resources in a multimodal text. Halliday (1985: 4) explains that linguistics is at the same time a "kind of semiotics" because language is viewed as "one among a number of systems of meaning that, taken all together, constitute human culture." In particular, technology has accentuated the multimodal nature of text, by facilitating the production and consumption of visual texts. Visual texts such as webpages have images, both static and dynamic, that work together with language to convey meaning multimodally. In addition, webpages may also include various audio and sound effects that, together with the interactive links, offer an intensely multimodal viewing experience not available from reading a printed book. The epistemological implication of multimodality is that meanings in a text can no longer be assumed to be the result of a single semiotic resource. Meanings are the result of the collective semiotic resources co-deployed within the same text. The multimodal approach takes into account how language and image (as well as other) choices fulfill the purposes of the text, the audience and context, and how those choices work together in the organization and development of information and ideas.

Within the primary and secondary education setting, Unsworth makes the following observation: "While many of the fundamentals of established, language-based literacy pedagogies will endure in the foreseeable future, they are by no means sufficient for the development of the kinds of literacy practices that already characterise the continuously evolving information age of the new millennium" (Unsworth 2002: 62).

Traditional literacy, that is the ability to use language competently for reading and writing, will retain its importance. However, with the increasingly multimodal nature of communication in this digital age (Kress 2003), it is important for us to develop the literacy to make sense of the new knowledge, to discern truths from falsehoods, and to evaluate the validity of these multimodal texts. In light of this, Kress (2003) proposes a shift from an alphabetic literacy to a multimodal literacy. He argues that this will facilitate changes in how literacy is developed in school. The need for multimodal literacy also grows proportionally more pressing as interactive digital media and information technology become even more ubiquitous (see e.g., Kalantzis, Cope and Harvey 2003 and Jewitt 2007).

More recently, there has been growing recognition (see e.g., Unsworth 2014; Unsworth and Macken-Horarik 2015; Chan and Chia 2014; O'Halloran and

Lim-Fei 2011; and Lim-Fei et al. 2015) that it is important to develop multimodal literacy in our students. Students need to develop the competencies to view multimodal texts critically. The challenge for educators is how to teach critical viewing and develop multimodal literacy in their students. Developing multimodal literacy involves acquiring the codified knowledge in the field, recognizing how it can be appropriated and transferred to new contexts, and reproducing this understanding through demonstrated competencies in performance tasks.

Multimodal literacy, first proposed by Jewitt and Kress (2003), is about understanding the different ways of knowledge representation and meaning making. Multimodal literacy "focuses on the design of discourse by investigating the contributions of specific semiotic resources (e.g., language, gesture, images) co-deployed across various modalities (e.g., visual, aural, somatic), as well as their interaction and integration in constructing a coherent text" (Lim-Fei et al. 2015: 917). Multimodal literacy aims to develop students into discerning readers and savvy producers of multimodal texts by drawing attention to the various strategies utilized in the production of these texts, and the ways in which specific choices work together to achieve the desired communicative goals. Current research in multimodal analysis establishes the need and provides the metalanguage to develop multimodal literacy in education. O'Halloran and Lim-Fei (2011: 14) envision that "a 'multimodal literate' student must be sensitized to the meaning potential and choices afforded in the production of the text, rendering an enhanced ability to make deliberate and effective choices in the construction and presentation of knowledge."

In the last decade, many frameworks and approaches have been developed to examine the meanings made in multimodal texts. For instance, in films (Bateman and Schmidt 2012; Bateman and Wildfeuer 2014; Tseng and Bateman 2010; Wildfeuer 2014), picture books (Painter 2008, 2013; Painter, Martin and Unsworth 2011; Wignell 2011), print advertisements (O'Halloran 2008; O'Halloran and Lim-Fei 2009), and television advertisements (Baldry and Thibault 2006; Feng and Wignell 2011; Lim-Fei and O'Halloran 2012). Advances in theoretical understandings in multimodality have also been compiled in recent publications such as O'Halloran and Smith (2011), Jewitt, Bezemer and O'Halloran (2016), and Bateman, Wildfeuer and Hiippala (2017).

Given the theoretical understandings developed in recent research in multimodality, it is worthwhile to explore how they can be extended to inform the teaching and learning of students in the secondary education setting. While many of the extant theories are meant for graduate and postgraduate research work, this chapter investigates how these understandings may be applied in the secondary education setting. It is important for teaching and learning at the secondary education level to be informed and grounded in sound theoretical understanding. Our students must be equipped with the skills and knowledge to comprehend the messages in the multimodal texts. They must learn to view such texts with discernment, recognize perspectives, and clarify their values in relation to these messages. Developing these critical viewing competencies requires a deliberate focus by the teacher to scaffold the students' viewing process, impart the language and tools to "deconstruct" or analyze the text, as well as cultivate the dispositions and attitudes toward these media texts.

However, it must also be recognized that there are constraints in the secondary education setting. These include a fairly crowded curriculum in most systems where time and space to be devoted to new areas of learning are strongly contested (Tan 2006). While teachers may have access to professional learning opportunities, the range of teachers' capabilities can be wide and uneven in many systems. As such, teachers may be unwilling or unable to manage new knowledge that may appear too technical and challenging to appropriate (Albright and Kramer-Dahl 2009; Teo 2014). Explicit alignment with current content areas of learning is also important, as teachers have to make connections and create coherence across the disparate areas of learning for their students (English Language Syllabus 2010 [Primary and Secondary]: 16). In this light, it can be challenging to apply the current theories and frameworks in multimodality directly to inform teaching and learning in the classroom. There is, therefore, a need for a translational research process, where the insights from research can be meaningfully adapted and approximated to inform the practices in the secondary education setting. The aim is to develop a theoretically robust yet easily accessible framework to scaffold the analysis of multimodal texts for the young and teach critical viewing.

3 The systemic approach to critical viewing

The English Language Syllabus guides the teaching of English Language literacy in Singapore. In the revised syllabus from 2010, it is mentioned that this English language curriculum will be

> enriched through the use of a variety of print and non-print resources that provides authentic contexts for incorporating the development of information, media and visual literacy skills in the teaching of listening, reading, viewing, speaking, writing, and representing [as well as] opportunities for pupils to be

> exposed to and engage in producing a variety of multimodal texts to represent ideas effectively and with impact. (English Language Syllabus 2010 [Primary and Secondary]: 9)

With this, two new areas for language learning, namely Viewing and Representing, are added. As part of the teaching of viewing skills, teachers are expected to help students "comprehend closely and critically a variety of different types of texts: literary and informational/functional, print and non-print [and] teach pupils to think critically and reflect on what they read and/or view to become critical readers and viewers" (English Language Syllabus 2010 [Primary & Secondary]: 29). With the inclusion of the aim of developing students into critical viewers being reflected in the English Language Syllabus in Singapore, it is a signal that the notion of literacy within the primary and secondary education setting in Singapore schools has broadened beyond the traditional areas of language learning, such as reading, writing, and speaking, to a literacy that takes into account the multimodal communicative environment the students inhabit. It recognizes the importance of fostering multimodal literacy among students, starting as young as when they are in primary school, that is the age of seven in Singapore.

While the acknowledgment of the importance of multimodal literacy and having it represented in the Singapore English Language Curriculum signifies good progress, there remains, understandably, a policy intent and implementation gap. This is unsurprising since, given the introduction of the new areas of language learning, teachers would need time to develop the knowledge and skills to be able to teach them. Time is needed for the professional competencies of the English language teachers to be built so that they, in turn, can nurture and develop their pupils into critical viewers.

The syllabus document indicates that "viewing skills will be taught explicitly" (English Language Syllabus 2010 [Primary & Secondary]: 20). For instance, teachers will guide students to "evaluate the logic and soundness of arguments by posing a range of questions [and evaluate] the validity of an argument based on the given evidence and the lines of reasoning presented" (English Language Syllabus 2010 [Primary & Secondary]: 22). Teachers are, thus, expected to guide students through the process of viewing and scaffold their learning such that they are able to develop their students into critical viewers. The challenge for teachers remains in knowing how to do so effectively.

From the authors' classroom observations, most teachers would typically "teach" visual texts by asking a series of questions, often framed randomly, to the students (Lim-Fei et al. 2015). The approach of "teaching" visual texts by questioning or "interrogating the text" is unproductive, as it assumes that by "testing" the students' comprehension through a barrage of questions, the understanding of the visual text will somehow develop. It assumes that the understanding of the text will be intuitively transferred to understanding of other texts. It assumes that through the experience, students will somehow develop into critical viewers. While a better-defined approach to teaching visual texts is not yet common in schools, attempts have been made in this direction. An example is the development of a guided set of questions or frames by Chan and Chia (2014), known as the Six Semiotic Modes Framework. Building on the earlier work by Anstey and Bull (2010), the framework provides descriptions and examples of the linguistic, audio, spatial, oral, visual, and gestural modes to guide the teachers and students as they make various observations when analyzing a multimodal text.

Using similar social semiotics lens as Chan and Chia (2014), this chapter proposes a genre-based pedagogy to the teaching of visual texts. Following the work of researchers such as Tan, Marissa and O'Halloran (2012), the authors adapted the work in the field of multimodality to develop a Systemic Approach and a FAMILY Framework to develop the knowledge and skills of secondary students to talk about, understand, and question the meanings made in multimodal texts.

The Systemic Approach to teaching visual texts emphasizes explicit teaching of the generic features of visual texts and introduces the common multimodal strategies used to engage viewers. It aims to provide a set of pedagogical scaffolds, informed by Systemic-Functional Theory and insights from multimodal research, to provide students with a structure and appropriate metalanguage to interpret visual texts. The Systemic Approach is named as such, given its roots in Systemic-Functional Multimodal Discourse Analysis (SFMDA) (Jewitt, Bezemer and O'Halloran 2016; O'Halloran and Lim-Fei 2014), which is an application of the Systemic-Functional Theory developed by Halliday (1985). As O'Halloran and Lim-Fei observe, "The term 'systemic' also describes the underlying organization of semiotic resources which enable the resources to be used for different purposes. The systems of meaning are typically modeled as inter-related 'system networks' (Halliday and Matthiessen 2004; Martin 1992; Kress and van Leeuwen, (2006) to describe the meaning potentials of semiotic resources" (O'Halloran and Lim-Fei 2014: 138).

Systemic-Functional Theory is about meaning as choice. Halliday explains that "systemic theory is a theory of meaning as choice, by which language, or any other semiotic system, is interpreted as networks of interlocking options" (Halliday 1994: xiv). Meaning is, therefore, made through realized choices from paradigms and in syntagms. Semiotic resources comprise networks of interlocking options from which the meaning maker selects. As Halliday explains, the choice is "not a conscious decision made in real time but a set of possible alternatives" (Halliday 1994: xiv–xxvi) from which choices are made in actual texts. The paradigmatic and syntagmatic options available in the system network foreground the importance of choice in Systemic-Functional Theory. As such, Lim-Fei notes: "The perspective offered by Systemic Functional Theory, and by extension SFMDA, is that meaning making is a result of choice. These choices may not always be conscious or intentional but they are always motivated according to the interest of the meaning maker" (Lim-Fei 2011: 74).

The Systemic Approach also offers a vocabulary to describe multimodal texts. It is important to have a metalanguage to denote semiotic resources beyond language so as to "describe meaning in various realms" (New London Group 2004: 24). Unsworth argues thus:

> Teachers and students need this kind of metalanguage for talking about language, images, sound, and so forth, and for their meaning-making interactions . . . . This kind of metalanguage gives students and teachers a means of comparing texts, of determining what semiotic choices were made in constructing particular meanings, what alternatives might have been chosen, and the effects of particular choices rather than others. (Unsworth 2014: 38)

With a metalanguage to describe the choices made in visual texts, students are able to identify the common media strategies used to engage them and the typical effects they bring about (Painter, Martin and Unsworth 2013; Unsworth and Cleirigh 2009). This leads to heightened awareness of the meanings in the visual texts. With the metalanguage, students are able to identify the genre of the texts, specifically the features and their typical functions, so that this knowledge will guide their reading of new texts within the genre that they will encounter, as has been well recognized (Martin 2012; Rose and Martin 2012). While a set of metalanguage, undergirded by sound theories, is useful, care must be taken not to overwhelm the teachers and students with too much technical jargon and complexities. As such, the translational process is critical. The authors worked iteratively with teachers to judiciously identify the necessary descriptions and choice of descriptors that are aligned with what they are already using to teach similar concepts in English language learning.

The metalanguage provided to the students empowers them to describe and discuss the visual texts. The Systemic Approach in teaching critical viewing focuses on the explicit teaching of features and strategies in a visual text supported by a framework to scaffold teaching and learning. As part of scaffolding the understanding of the visual texts, students are explicitly taught the generic features and typical functions of the texts. They are also introduced to the common strategies and typical effects used to engage the viewer. This equips the students with the understanding to know where to look and what to look out for in a visual text.

Through this procedure, students develop critical thinking and discourse analysis skills—which nurtures them into critical viewers, armed with the knowledge of text types and the common multimodal engagement strategies used in these texts. This approach is very much unlike how students make sense of a poem or prose in a literary criticism class in the subject of English literature. Within the Systemic Approach, the extension of discourse analysis is made beyond language to multimodal discourse analysis. Teachers who understand the multimodal ways in which knowledge is presented may teach students to assess, appraise, and appropriate the multimodal texts that they will inevitably encounter. In addition, through the lessons, students will become more discerning viewers of multimodal texts—thus developing multimodal literacy.

While questioning is still encouraged, the Systemic Approach advocates first the explicit teaching and identification of generic features and engagement strategies in the multimodal texts. Questions are used to elicit interpretations of the visual texts that are supported by textual evidence. Students are able to respond to these questions based on their knowledge of the textual features and typical functions as well as the common multimodal strategies and typical effects that have been introduced earlier. As students apply their knowledge of the text to make meaning from the visual texts, their critical reasoning faculties and discourse analysis skills develop. This is because they are now better equipped with the relevant skills to explain the choices made in the text and subsequently to present an argument for their interpretation.

The Systemic Approach consists of three levels of viewing with a visual text (see Table 8.1). They are:

Table 8.1 Three levels of viewing a visual text

LEVEL 1	**ENCOUNTER** Engaging with the text
LEVEL 2	**COMPREHENSION** Understanding the text
LEVEL 3	**CRITICAL VIEWING** Questioning the text

3.1 Level 1: Encounter—Engaging with the Text

The first level of "Encounter: Engaging with the Text" focuses on the affective domain where students react to the text based on their immediate impressions. The Systemic Approach encourages teachers to devote time and space during the lesson for students to engage, on an emotional level, with the text.

Upon presenting a visual text, teachers can invite students to share personal responses to it. The intent is to address the affective component of students' engagement with the visual text. The responses may be individual or collective, written or oral, and may comprise emotional reactions, associations called forth by the text, and evaluations based on initial observations. These reactions can include finding the visual text humorous, boring, intriguing, or confusing. The reactions of students may also vary from indifference to deep personal involvement, and each reaction to the text will likely be based on a different initial observation of the text. Students are then invited to articulate the reasons (why) for their emotional reaction. In doing so, the students may express their focal awareness and associations that the text calls forth (and thus begin their first step toward critical viewing!). Students may point out a particular visual element that captures their attention, and what is visually dominant in a text may vary from student to student. For instance, a student may articulate attention that is riveted upon the face of the protagonist and is acutely aware of the latter's delight. Another student may express attention to the items that are presented in vivid colors against a dull background. Another student may discuss personal associations, feelings, and ideas engendered by the viewing. As students seek to make sense of the text, they may recall objects, people, and events in their lives. Some may be reminded of other texts—they could be visual, aural, or both—of a similar theme but composed of different visual and linguistic elements. Others may be reminded of their past experience(s); for instance, they may feel more intensely toward an antismoking advertisement because they had a family member whose terminal illness was caused in part by heavy smoking. All of these show a powerful resonance of personal observation and experience in the interaction with a visual text. The teacher consolidates students' responses, invites expansion wherever appropriate, and concludes by drawing attention to different aspects of the multimodal text and the underlying differences in understanding that account for varying responses. In this way, the teacher helps students become more aware of the need to examine and integrate all the elements in the text (linguistic, visual, and others) so as to arrive at an informed and coherent interpretation of the multimodal text. This is the first step to active engagement with the visual text and its ideas.

3.2 Level 2: Comprehension—Understanding the Text

The understanding of the visual texts must be explicitly taught and the learning of it well scaffolded. "Level 2: Comprehension—Understanding the Text" anchors the Systemic Approach as this is the level that adapts the theories and frameworks in multimodality into an accessible framework to guide the students' critical viewing of visual texts (see Figure 8.1).

As discussed earlier, the Systemic Approach provides students with a language to describe multimodal texts. It also helps students develop awareness of the genre of the visual text, that is, of the text's features and the typical functions of these features. Finally, the Systemic Approach leads students to acquire a

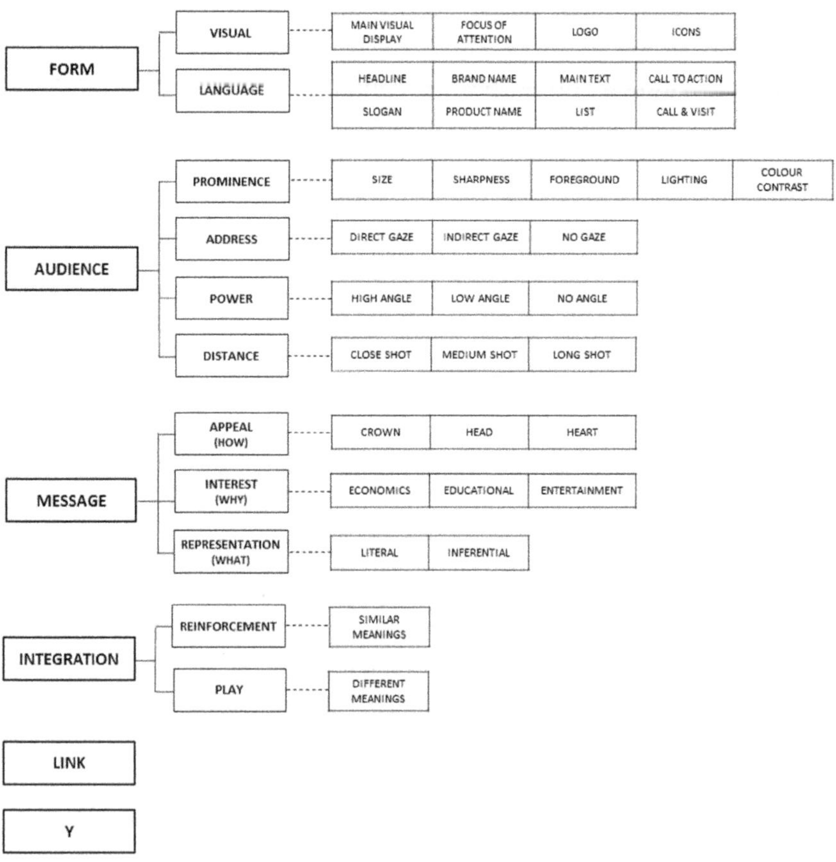

Figure 8.1 FAMILY framework.

sensitivity to the common multimodal strategies and their typical effects. As such, the Systemic Approach scaffolds the understanding of the multimodal texts through explicating the system choices and the ideational, interpersonal, and textual metafunctions they realize. This follows from the understandings in Systemic-Functional Theory, where the meanings made in language through the system choices oriented around the ideational, interpersonal, and textual metafunctions (Halliday 1994; Halliday and Matthiessen 2004).

The aims of the Systemic Approach are realized through the FAMILY Framework that has been developed through working alongside teachers (see Figure 8.2). The FAMILY Framework is a result of distilling the understandings from the SFMDA approach (Jewitt, Bezemer and O'Halloran 2016; O'Halloran and Lim-Fei 2014) and representing the ideas in an easy and accessible manner for teachers and students in the secondary education setting. For instance, the textual metafunction is represented as Form, the interpersonal metafunction as Audience, and the ideational metafunction as Message. While retaining some of the core ideas of SFMDA, the FAMILY Framework is also eclectic in the sense that it draws on insights from the field of media studies and rhetorical studies, for instance, the Aristotelian types of persuasion in ethos, logos, or pathos (see, e.g., Halmari and Virtanen 2005 and Ross 2010).

FORM

Acquiring awareness of the **genre** (**features** and **functions**)

AUDIENCE

Developing sensitivity to the common multimodal **strategies** and their typical **effects**

MESSAGE

Understanding literal and inferential **representations** and nature of **persuasion**

INTEGRATION

Recognising **relationship** across **semiotic modes**

LINK

Connecting across **areas of language learning** and other **subjects**

Y

Questioning to prompt **evidence-based interpretation**

Figure 8.2 Summary of systemic approach.

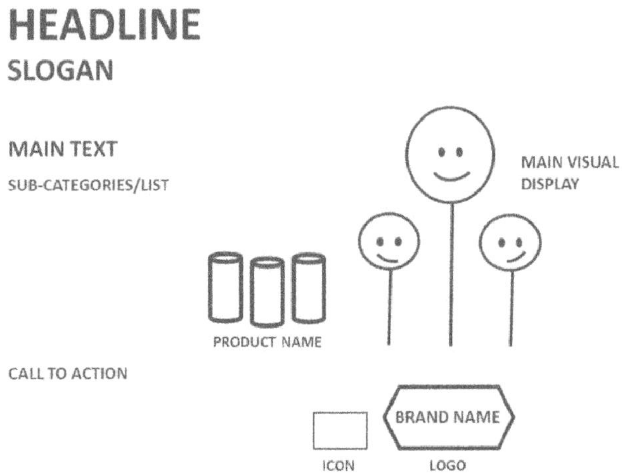

Figure 8.3 Parts of a typical visual text.

Form

Adapting from the systems under the textual metafunction by Tan, Marissa and O'Halloran (2012), students first learn the parts of a visual text and relate the parts in the text to the typical functions they serve (*FORM*). The information that is offered to viewers is typically subsumed into typical features in an advertisement text. These are the headline, slogan, main text, subcategories or list, main visual display, call to action, icon(s), and logo (see Figure 8.3).

Students look at the individual layers of information to develop a preliminary understanding of the text. Table 8.2 provides a summary of the parts of a visual text and the typical functions they serve.

Students learn that most visual texts have a headline that is typically short and punchy and can be placed anywhere in the visual text. Usually large, set in boldface, or a contrasting type or color, it can be identified by its appearance. This attracts viewer attention and arouses interest toward the rest of the visual text with the topic expressed or implicit in the headline. A slogan is a catchy phrase that captures viewer attention and conveys the brand's key selling proposition in a compelling and memorable way. To make a slogan memorable, some writers use devices such as alliteration (e.g., Jaguar's "Don't dream it. Drive it."), pun (Citibank's "Because the Citi never sleeps"), and rhyme (Pringles' "Once you pop, you can't stop"). The main text portion, produced in smaller font, contains details and descriptions; it explains ideas presented

Table 8.2 Description of parts and typical function(s) served (adapted from Tan, E, and O'Halloran 2012)

Form	Description	Typical function(s)
Visual		
Main Visual Display	Largest and most prominent visual	To attract viewer attention and arouse interest
Focus of Attention	Most salient visual feature(s)	To focus viewer attention on what is important
Logo	Graphic representation of organization	For identification of organization
Icon	Graphic representation of ideas	For faster information processing
Language		
Headline	Largest and most prominent text	To attract viewer attention and arouse interest; to express or imply topic
Slogan	Catchy promotional phrase	To convey brand's key selling proposition
Main Text	Smaller text comprising main information	To provide more details and description
Brand Name	Name of brand, company	For brand identification or awareness
Product Name	Name of product	For product identification or awareness
Call to Action	Command to do something	To solicit viewer to take action

in the headline and other elements of the visual text. Sometimes, subheadings and a bulleted list are used to ease reading. Other elements include the brand name and product name that aid the viewer in brand and product identification and awareness. A call to action is typically placed at the bottom of the visual text. Reinforcing the messages of the headline and supporting text, it strongly solicits the viewer to take some action. Varying with the nature of the visual text, it includes imperative verbs such as "call," "write," "try," "visit," "email," "order," or "buy." Some visual texts also invite viewers to scan quick response (QR) codes provided for easy access to the company's website, buy eCoupons, or use coupon codes.

Having looked at the linguistic elements of a visual text, students will then examine the illustrations. The dominant image, which can be referred to as the main visual display, captures viewer attention. The focus of attention then shifts to the most salient features of the illustration, which can be intentionally elicited with the use of techniques such as color contrasts and lighting. Other visuals include organization logos that are often displayed in the lower-right corner, which allow the organizations to be readily identified and create an obvious link between the organizations and the products, services, or ideas presented; and icons that are universally understood and, thus, allow for faster information processing as compared to that for the corresponding words.

Audience

Audience, adapted from the systems under the interpersonal metafunction by Tan, Marissa, and O'Halloran (2012), introduces students to the common strategies used in visual texts to attract attention. These help students develop an understanding of codes and conventions and their influence on the viewer.

Students learn how an element can be given prominence, or made to "stand out." This salience, as depicted in Figure 8.4a, can be realized by choices made in size, sharpness of focus, color contrast, lighting, and foreground techniques.

Students learn that the subjects' type of gaze—looking directly at or away from the viewer—changes the way the viewer interacts with them. In Figure 8.4b, the first subject is looking "out of the frame"; through this, the subject addresses the viewer and, typically, demands the latter to do something. The demand for connection is reinforced with the placing of viewers at eye level with the subject. This is contrasted with the gaze of the second subject who does not look directly at the viewer—and this establishes the subject as an "object" offered as an item of information for contemplation.

Following the work of scholars such as Kress and van Leeuwen (2006), students learn that camera distance (see Figure 8.5 and Table 8.3), that is, the space between the camera and its subject(s), can change the degree of emotional involvement the audience has with the subjects. In essence, medium shots and

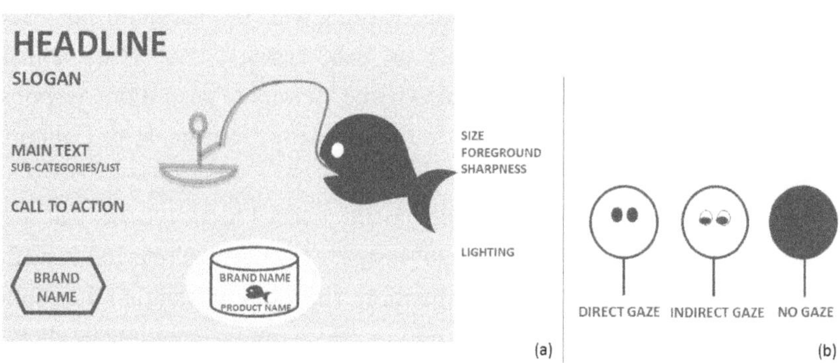

Figure 8.4 (a) Prominence; (b) Address.

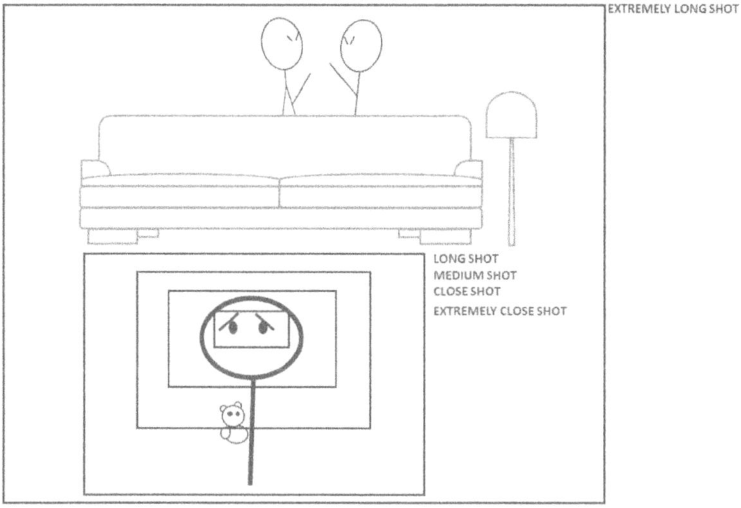

Figure 8.5 Shot types.

Table 8.3 Summary of shot types and their effects

Shot types	Description	Effect
Extreme long shot (XLS)	The human subject is very small in relation to the surrounding environment	Location
Long shot (LS)	The entire body of the subject is visible, along with some of the surrounding space	Context
Medium shot (MS)	The subject is framed from the waist up	Personal
Close-up (CU)	A section of the body is framed, for example, face, torso, hands	Intimacy
Extreme close-up (XCU)	Only a body part is framed, for example, eye, ear, finger	Scrutiny

close-ups tend to create a greater sense of intimacy by allowing viewers to focus on the subjects' faces and emotions, while long shots tend to accentuate the environment and the space surrounding the subjects.

Kress and van Leeuwen (2006) also posit that generally relationships of power are constructed through the use of vertical angles. High-angle shots, where the camera is positioned above the subject or action and aimed downward, tend to minimize the subject (see Figure 8.6a). The subject is rendered powerless and vulnerable. When the camera is positioned below the subject in low-angle shots and aimed upward, it amplifies the size and volume of the subject, rendering him or her more important, powerful, and imposing in relation to the viewer than if the camera angle is at eye level.

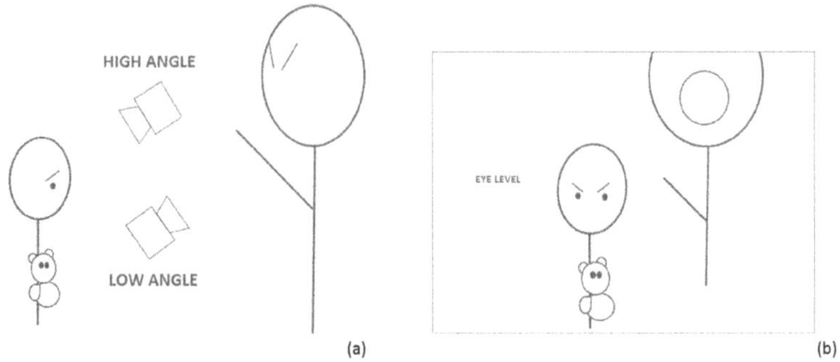

Figure 8.6 (a) Camera angle determines power relations; (b) An eye-level shot.

It should be noted, however, that more recent studies in multimodal film analysis, such as Bateman and Schmidt (2012) and Bateman and Wildfeuer (2014), have questioned the generalization of the use of vertical angles and power relations. While it must be recognized that there are examples where the opposite is true, it is nonetheless useful for students to know of the effects that are typical but not always realized by specific choices in camera angles. Through this, students build awareness of conventional interpretations of camera angles. It is, nevertheless, important to note that such conventional interpretations may be overridden by other factors in particular contexts. Take Figure 8.6b as an example, where the camera is placed at eye level with the child's head, and behind him, captures in extreme close-up his towering parent's reprimands. A possible interpretation is the child's assertion of power (and noncompliance) and significance in the text—with him turning his back to the figure of authority, seizing the viewer's gaze, and occupying the center of the frame. In this case, this camera angle, predicated on a convention where eye-level framing, either implies an equal power relationship between the subject and the viewer or is merely an alignment of the viewer's line of sight with the screen, creates a shot that is meaning-laden. It augments the effect of the power and importance of the subject. This is an example of shots that encourage viewers to look beyond default interpretations of camera positioning.

Message

Building on this understanding, students identify the type of persuasion (ethos, logos, and pathos are the three modes of persuasion) used to appeal to the viewer and discuss the literal and inferential meanings in the visual text (*MESSAGE*).

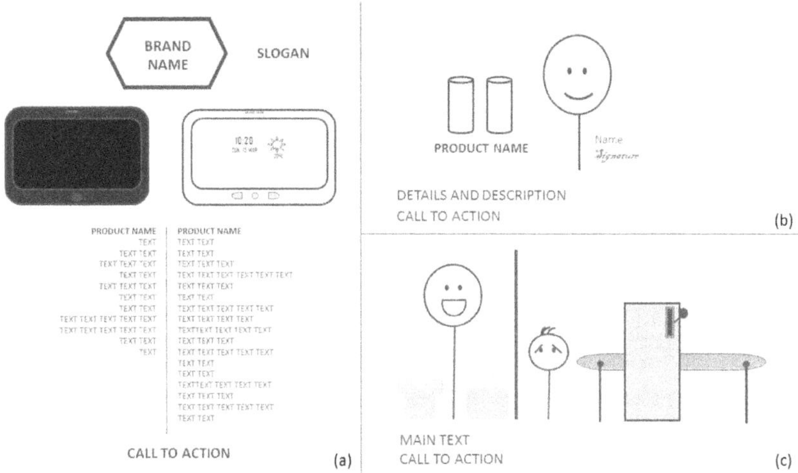

Figure 8.7 The nature of the persuasion used to appeal to the viewer: (a) Appeal to head (logos); (b) Appeal to crown (ethos); and (c) Appeal to heart (pathos).

Students learn the three types of persuasive *appeals*: (head) logos, (crown) ethos, and (heart) pathos. For logos, the mode of persuasion takes place on the rational level. "Ethos" is the appeal to the authority of the subject, and "pathos" is the appeal to the emotions of the audience. While there have been scholars, such as Fleming (1996) and Patterson (2010), who oppose the idea of images being a form of argument as these types of persuasions are originally verbal, others, such as Kjeldsen (2015), maintain that visual argumentation based on these types of appeals is relevant. It has also been noted by scholars, such as Bhatia (2005) and Bruthiaux (2005), however, that these types of appeals are seldom discrete and operating in isolation. For instance, Halmari and Virtanen (2005: 5–6) observe that "emotional appeals are also found, as expected, in the language of advertising, combined with ethos (linguistically mediated implications of the 'good character' of the persuader) and logos (the appeal to the rationality of the audience)." While recognizing that these types of persuasion can be combined in visual texts, it is, nonetheless, useful and important for students to know and be able to recognize them in a text. The three types of Aristotelian appeals are translated for teachers' and students' understanding below.

Head (logos, or appeal to reason) demonstrates an effective use of reason and judicious use of evidence such as facts, statistics, comparisons, professional opinions, anecdotes, or observations. Figure 8.7a is an example of how logos is employed: The juxtaposition of two devices, with one being more salient with its features and specifications to effect persuasion. *Crown* (ethos, or appeal to credibility) demonstrates goodwill toward the audience and the professional knowledge and experience of the subject at hand. Hence, professionals are selected to back up a cause or product; sometimes, companies select celebrities to endorse the product to ride on their authority of fame, glamour, or acclaimed expertise (see Figure 8.7b for a typical example). *Heart* (pathos, or appeal to emotion) entails the use of visuals and language that will connect with the feelings of the audience. Figure 8.7c is an example of how pathos is employed. The child, with a heart-rending expression on his face, engenders sympathy in the viewer as the latter realizes that he is a child laborer; the juxtaposition of the child with a delighted consumer raises awareness of the link between consumerism and child labor, and how consumer actions are and can become effective mechanisms of global change. As mentioned earlier, more than one mode of appeal may be employed in some visual texts.

Students learn that visual texts serve the interest(s) of the producer or organization: economics, education, and entertainment. The visual text in Figure 8.8 clearly serves an economic interest—to sell a service. There are also multimodal texts produced for educational or entertainment purposes. Students also learn that some texts may serve more than one interest; for instance, the use of entertainment or humor in a visual text intended for education increases positive affect, making it more compelling and memorable.

Figure 8.8 Literal and inferential meanings.

When engaging in meaning making, students also learn that visual texts can contain literal and inferential meanings. This is illustrated by Figure 8.8, where at the literal level, the company organizes children's birthday parties; and at the inferential level, the company appears to intimate that it organizes successful birthday parties filled with organized fun for both children and adults. With the provision of instructional scaffolding, students learn to draw on their prior knowledge and experience in their inferential interpretation of the text.

Integration

Finally, students examine the relationship between the language and the image in terms of similar (see Figures 8.9a and 8.9b) or different meanings made between the two modalities (see Figure 8.9c) (INTEGRATION). By the end, students understand the interaction between form and content in visual texts, and more specifically, how the integration of the linguistic and visual elements can lead to a coherent unified representation and thereby achievement of the intended purpose(s).

Such a structured approach for the analysis of form and content in multimodal texts helps students get beyond a purely subjective reaction to the text and become critical viewers who examine and synthesize print and visual information to formulate informed interpretations.

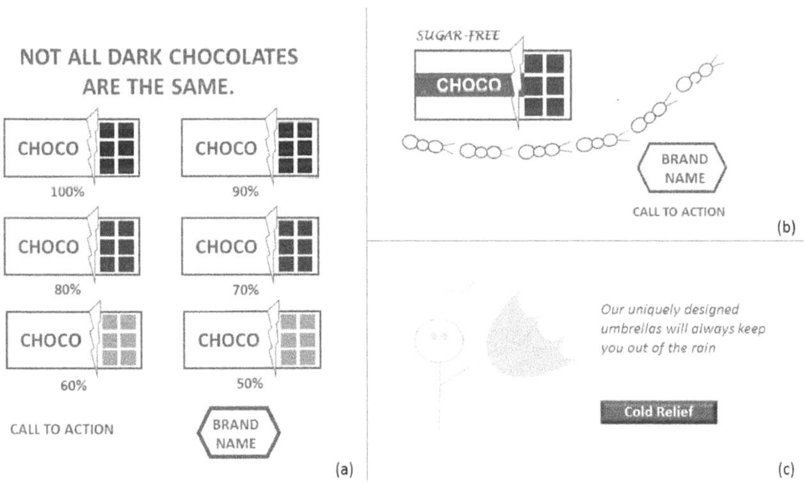

Figure 8.9 Relationship between the language and the image: (a) Visual and linguistic partnership; (b) Reinforcement of message through the use of irony; and (c) Different messages conveyed by the visual and the linguistic.

Link

Beyond *Form, Audience, Message*, and *Integration*, the following *Link* and *Y* are essentially reminders for the teachers as they teach visual texts. The authors, on several occasions, have ironically confessed that they are also, in part, our attempt to form a sensible acronym that can serve as a mnemonic device for teachers. *Link* refers to the importance for teachers to align what students are learning within the FAMILY Framework with other areas of language learning as well as, possibly, other subjects. For instance, in terms of other areas of language learning, in *Message*, the strategies of identifying literal and inferential meanings are similar for visual texts and for verbal reading skills. Likewise, in terms of extension to other subjects, in *Message*, the types of persuasions are also found in texts that students may encounter in science and the humanities as well.

Y

The *Y* in the FAMILY Framework stands for questioning. As discussed earlier, a common approach to teach critical viewing among teachers is through questioning. The role of questions is also represented in the FAMILY Framework. However, instead of being the only strategy for teaching visual texts, it is represented as one of the strategies used within the FAMILY Framework. The role of questioning here is to prompt evidence-based interpretations from the students. Through the use of questioning, teachers can guide students in making explicit the students' tacit understanding of the visual texts. Having developed a certain understanding of the genre of visual texts and the common engagement strategies used in these texts, students are now prepared to cite textual evidence to support their responses to the text. While questioning is still used as a strategy, the scaffolds provided in the FAMILY Framework contrast with the previous approach where students can only make intuitive responses based on their personal experiences to the questions asked. As the teacher asks questions to elicit students' responses to the text, students are now equipped with the vocabulary and the knowledge, through the scaffolds provided, to interpret the text and defend their views with textual evidence.

3.3 Level 3: Critical Viewing—Questioning the Text

It is evident that Level 2 "Comprehending the Text" requires a slightly higher level of multimodal text processing than Level 1. It promotes critical thinking by equipping students with a metalanguage to describe the text as well as the strategies and skills essential to unpacking the layers of meaning in the text.

Level 3 "Questioning the Text" is designed to support, challenge, and encourage students to further develop their critical thinking and reasoning skills. It requires the highest level of processing. At this level, viewers evaluate the effectiveness of the different semiotic resources deployed to fulfill certain purposes. In other words, students critique the multimodal text in terms of content, design, and cohesion. The teacher can pose questions such as:

- Do the text and image(s) converge to engender powerful, persuasive messages? Why?
- What can be done to make the piece work (even) more effectively for the intended audience? Why?

These questions can be broken down further to scaffold students' reasoning processes: Is the message clear? How persuasive or informative is the text? Do the language and image(s) work together to effectively convey the message(s)? Are there images or words that should be removed, added, or modified? These are some example questions, but teachers are not limited to them. For instance, for a text that uses pathos, an appeal to emotion, the teacher may ask: Does the text create that emotional connection with the viewer effectively? Each of these questions should be followed by the question "why" to prompt thoughtful, well-reasoned arguments. Such questions can stimulate students to action, to do something with the information acquired or perspectives developed; and teachers can provide guidance pertaining to the types of products that can demonstrate deep understanding. Student responses can be written or can take some other creative forms. Students may, individually or collectively, modify a multimodal text or compose one situated within the original or an entirely new context and articulate how their texts deploy and interrelate the resources of language and image (and other semiotic modes) to affect persuasion. Teachers can engage students in discussions on expected proficiency and indicators of quality and provide ongoing formative feedback—peer feedback may also be employed—to help students improve their work.

4 Conclusion

There remains a perennial need for teaching and learning in the classroom to be informed by the best of research understandings. Applications of research into the classroom allow for a certain utility for the theories and frameworks developed as well as a certain rigor in teaching and learning. As such, translational research plays a vital role in bridging empirical and theoretical developments with the needs of the classroom.

As the new digital world demands new literacies from our students, our curriculum needs to be continuously updated to ensure that the knowledge and competencies our students develop remain relevant. Teachers also need to expand their professional repertoire, to develop strategies in teaching new areas of knowledge, as well as to foster new areas of competencies in their students. As discussed in this chapter, our students need to develop multimodal literacy—to understand and evaluate visual texts critically. This is a new area of knowledge that requires teachers to draw on the latest developments in multimodality research and adapt them for the learning needs of their students.

The Systemic Approach for the teaching of visual texts has been iteratively refined as it was used in secondary schools in Singapore over the last few years. Most teachers have used the Systemic Approach in the teaching of students at the lower secondary school level. The Systemic Approach can help teachers facilitate the development of these learning outcomes, including critical thinking. In addition, through authentic learning experiences situated in real-life contexts, students become more discerning viewers of multimodal texts—thus acquiring multimodal literacy.

Future development of the Systemic Approach involves adapting the FAMILY Framework to teach students at the primary school level. This involves further simplifying the language used and making the framework accessible for younger students. More importantly, lesson resources and exemplars will be developed to illustrate how the Systemic Approach can be meaningfully applied in the classroom. Work has also begun recently in extending the Systemic Approach for the teaching of film texts. A FAMILY Framework for the teaching of film texts, incorporating both the visual and aural modes, is being developed and trialed in a secondary school in Singapore. This builds on and extends from the work of the present Systemic Approach and FAMILY framework. When completed, it is envisioned that both frameworks will be complementary and productive as scaffolds in developing multimodal literacy for primary and secondary school students.

References

Albright, J. and A. Kramer-Dahl (2009), "The Legacy of Instrumentality in Policy and Pedagogy in the Teaching of English: The Case of Singapore," *Research Papers in Education*, 24 (2): 201–22.

Baldry, A. and P. J. Thibault (2006), *Multimodal Transcription and Text Analysis: A Multimedia Toolkit and Coursebook*, London: Equinox.

Bateman, J. A. and K.-H. Schmidt (2012), *Multimodal Film Analysis: How Films Mean*, New York: Routledge.

Bateman, J. A. and J. Wildfeuer (2014), "A Multimodal Discourse Theory of Visual Narrative," *Journal of Pragmatics*, 74: 180–208.

Bateman, J. A., J. Wildfeuer, and T. Hiippala (2017), *Multimodality: Foundations, Research and Analysis – A Problem-Oriented Introduction*, Berlin: De Gruyter Mouton.

Bhatia, V. K. (2005), " Generic Patterns in Promotional Discourse," in H. Halmari and T. Virtanen (eds), *Persuasion across Genres: A Linguistic Approach*, 213–25, Amsterdam: John Benjamins.

Bruthiaux, P. (2005), "In a Nutshell: Persuasion in the Spatially Constrained Language of Advertising," in H. Halmari and T. Virtanen (eds.), *Persuasion across Genres: A Linguistic Approach*, 135–51, Amsterdam: John Benjamins.

Bull, G. and M. Anstey (2010), *Evolving Pedagogies: Reading and Writing in a Multimodal World*, Sydney: Education Services Australia Limited.

Chan, C. and A. Chia (2014), *Reading in the 21st Century: Understanding Multimodal Texts and Developing Multiliteracy Skills*, Singapore: McGraw-Hill.

Feng, D. and P. Wignell (2011), "Intertextual Voices and Engagement in TV Advertisements" *Visual communication*, 10 (4): 565–88.

Fleming, D. (1996), "Can Pictures be Arguments?" *Argumentation and Advocacy*, 33 (1): 11–22.

Halliday, M. A. K. (1985), "Part A," in M. A. K. Halliday and R. Hasan (eds.), *Language, Context, and Text: Aspects of Language in a Social-Semiotic Perspective*, 1–49, Geelong: Deakin University Press.

Halliday, M. A. K. (1994), *An Introduction to Functional Grammar*, 2nd edition, London: Hodder Education.

Halliday, M. A. K. and C. M. I. M. Matthiessen (2004), *An Introduction to Functional Grammar*, 3rd edition, London: Hodder Education.

Halmari, H. and T. Virtanen (2005), " Persuasion across Genres: Emerging Perspectives," *Pragmatics & Beyond*, 130: 3–24.

Jewitt, C. (2007), "Multimodality and Literacy in School Classrooms," *Review of Research in Education*, 32 (1): 241–67.

Jewitt, C. and G. Kress (eds.) (2003), *Multimodal Literacy*. New York: Peter Lang.

Jewitt, C., J. Bezemer and K. L. O'Halloran (2016), *Introducing Multimodality*, London: Routledge.

Kalantzis, M., B. Cope and A. Harvey (2003), "Assessing Multiliteracies and the New Basics," *Assessment in Education: Principles, Policy & Practice*, 10 (1): 15–26.

Kjeldsen, J. E. (2015), "The Study of Visual and Multimodal Argumentation," *Argumentation: An International Journal on Reasoning*, 29 (2): 115–32.

Kress, G. (2003), *Literacy in the New Media Age*, London: Routledge.

Kress, G. and T. van Leeuwen (2006), *Reading Images: The Grammar of Visual Design*, 2nd edition, London: Routledge.

Lim-Fei, V. (2011), "A Systemic Functional Multimodal Discourse Analysis Approach to Pedagogic Discourse," PhD diss., National University of Singapore, Singapore.

Lim-Fei, V. and K. L. O'Halloran (2012), "The Ideal Teacher: Analysis of a Teacher-Recruitment Advertisement," *Semiotica*, 189 (1/4): 229–53.

Lim-Fei, V., K. L. O'Halloran, S. Tan, and K. L. E. Marissa (2015), "Teaching Visual Texts with Multimodal Analysis Software," *Educational Technology Research and Development*, 63(6), 915–35.

Martin, J. R. (2012), *Genre Studies Vol. 3: Collected Works of J R Martin*, Shanghai: Shanghai Jiao Tong University Press.

Martin, J. R. and D. Rose (2012), *Learning to Write, Reading to Learn: Genre, Knowledge and Pedagogy in the Sydney School*, Sheffield, UK: Equinox.

Matthiessen, C. M. I. M. (2013), "Appliable Discourse Analysis," in F. Yan and J. J. Webster (eds.), *Developing Systemic Functional Linguistics: Theory and Application*, 138–208, London: Equinox.

Ministry of Education (2010), *English Language Syllabus 2010: Primary & Secondary (Express/Normal [Academic])*, Singapore: Curriculum Planning & Development Division, Ministry of Education.

New London Group (2000), "A Pedagogy of Multiliteracies: Designing Social Futures," in B. Cope and M. Kalantzis (eds.), *Multiliteracies: Literacy Learning and the Design of Social Futures*, 9–38, Melbourne: Macmillan.

O'Halloran, K. L. (2008), "Systemic Functional-Multimodal Discourse Analysis (SF-MDA): Constructing Ideational Meaning Using Language and Visual Imagery," *Visual Communication*, 7 (4): 443–75.

O'Halloran, K. L. (2011), "Multimodal Discourse Analysis," in K. Hyland and B. Paltridge (eds.), *Companion to Discourse*, 120–37, London: Continuum.

O'Halloran, K. L. and V. Lim-Fei (2011), "Dimensioner af Multimodal Literacy," *Viden om Læsning* 10, September 2011: 14–21, Copenhagen, Denmark: Nationalt Videncenter for Laesning.

O'Halloran, K. L. and B. A. Smith (eds.) (2011), *Multimodal Studies: Exploring Issues and Domains*, New York: Routledge.

O'Halloran, K. L. and V. Lim-Fei (2014), "Systemic Functional Multimodal Discourse Analysis," in S. Norris. and C. D. Maier (eds.), *Texts, Images and Interactions: A Reader in Multimodality*, 137–54, Boston: De Gruyter.

O'Toole, M. (2010), *The Language of Displayed Art*, 2nd edition, London: Routledge.

Painter, C. (2008), "The Role of Colour in Children's Picture Books," in L. Unsworth (ed.), *New Literacies and the English Curriculum*, 89–111, London: Continuum.

Painter, C. (2013), *Reading Visual Narratives: Inter-Image Analysis of Children's Picture Books*, London: Equinox.

Painter, C., J. R. Martin and L. Unsworth (2011), "Organizing Visual Meaning: Framing and Balance in Picture-Book Images," in S. Dreyfus, S. Hood, and M. Stenglin (eds.), *Semiotic Margins: Meanings in Multimodalities*, 125–43, London: Continuum.

Painter, C., J. R. Martin, and L. Unsworth (2013), *Reading Visual Narratives: Image Analysis of Children's Picture Books*, Sheffield, UK: Equinox.

Patterson, S. W. (2010), "A Picture Held Us Captive: The Later Wittgenstein on Visual Argumentation," *Cogency*, 2 (2): 105–34.

Ross, W. D. (2010), *Rhetoric by Aristotle*, New York: Cosimo Inc.

Tan, C. (2006), "Creating Thinking Schools through 'Knowledge and Inquiry': The Curriculum Challenges for Singapore," *The Curriculum Journal*, 17 (1): 89–105.

Tan, S., K. L. E. Marissa, and K. L. O'Halloran (2012), *Multimodal Analysis Image (Teacher edition and student edition)*, Singapore: Multimodal Analysis Company.

Teo, P. (2014), "Making the Familiar Strange and the Strange Familiar: A Project for Teaching Critical Reading and Writing" *Language and Education*, 28 (6): 539–51.

Tseng, C. and J. A. Bateman (2010), "Chain and Choice in Filmic Narrative: An Analysis of Multimodal Narrative Construction in 'The Fountain,'" in C. R. Hoffmann (ed.), *Narrative Revisited: Telling a Story in the Age of New Media*, 213–44, Amsterdam: John Benjamins.

Unsworth, L. (2002), "Changing Dimensions of School Literacies," *Australian Journal of Language and Literacy*, 25 (1): 62–79.

Unsworth, L. (2006), "Towards a Metalanguage for Multiliteracies Education: Describing the Meaning-Making Resources of Language-Image Interaction," *English Teaching: Practice and Critique*, 5 (1): 55 76.

Unsworth, L. (2014), "Multiliteracies and Metalanguage: Describing Image/Text Relations as a Resource for Negotiating Multimodal Texts," in J. Coiro, M. Knobel, C. Lankshear, and D. Leu. (eds.) *Handbook of Research on New Literacies*, 379–409, New York: Routledge.

Unsworth, L. and C. Cleirigh (2009), "Multimodality and Reading: The Construction of Meaning Through Image-Text Interaction," in C. Jewitt (ed.), *The Routledge Handbook of Multimodal Analysis*, 151–63, London: Routledge.

Unsworth, L. and M. Macken-Horarik (2015), "Interpretive Responses to Images in Picture Books by Primary and Secondary School Students: Exploring Curriculum Expectations of a 'Visual Grammatics,'" *English in Education*, 49 (1): 56–79.

Wignell, P. (2011), "Picture Books for Young Children of Different Ages: The Changing Relationships between Images and Words," in K. L. O'Halloran and B. A. Smith (eds.), *Multimodal Studies: Issues and Domains*, 202–19, London: Routledge.

Wildfeuer, J. (2014), *Film Discourse Interpretation: Towards a New Paradigm for Multimodal Film Analysis*, New York: Routledge.

9

"Wikiganda": Detecting Bias in Multimodal Wikipedia Entries

Hartmut Wessler, Christoph Kilian Theil, Heiner Stuckenschmidt, Angelika Storrer, and Marc Debus

1 Introduction

Allegations of bias or the lack of neutrality are frequent in political and media realms, but also in everyday communication. The media portrayal of a particular politician is often labeled unfair or biased when that person or his or her supporters feel that important information has been omitted or the available information has been presented in unfavorable terms or with pejorative connotations. However, such allegations are hard to evaluate in an absolute sense: What is the universe of information that could and should have been included? And what would a perfectly neutral presentation of such information look like? For both the selection problem and the presentation problem, explicit standards are needed, which facilitate a valid judgment on the degree of bias or neutrality found in particular media offerings. This need is even more pressing when the analysis deals, as it does in this chapter, with a type of offering that is explicitly geared toward providing unbiased, factual accounts rather than interpretation, analysis, or opinion. Encyclopedias are meant to provide solid ground for everybody to work with, and in this regard it makes no difference whether the encyclopedia is written by selected experts or collaboratively compiled by its users, like in the case of Wikipedia.

In addition, the explorations presented here are set apart from most extant neutrality and bias research by focusing on multimodal documents. Written text and images are both used in detecting bias, as are their combinations, for example in the form of image-caption clusters or in the simultaneous analysis of visuals and text passages that anchor these visuals in the accompanying text. The aim of this contribution is to present a framework and empirical examples that show how different multimodal forms of communication such as texts, figures, and pictures can be used to send biased signals to the viewer or reader or to underline a specific position mentioned in the respective article. In general, pictures can be just as biased as written text, depending on their visual content and form (Moriarty and Popovich 1991; Verser and Wicks 2006; see also Debus, Stuckenschmidt, and Wessler 2015). And captions as well as anchoring passages can be used to enhance that bias or to attenuate it in addition to simply disambiguating the pictorial content (Martinec and Salway 2005).

Such relations between images and text segments can be studied on the micro level of individual articles or items. But such studies can also be complemented by macro-level research on typical patterns of text-image combinations in larger bodies of media content. Collections of multimodal documents such as news outlets, online encyclopedias, discussion forums, or media content repositories typically contain a limited number of such "multimodal frames," that is, typical combinations of aspects mentioned in text and aspects of pictorial content that in conjunction provide a distinct perspective on an issue, event, or actor across many different items (Wessler et al. 2016). Both micro- and macro-level investigations of how distinct perspectives and bias are created in multimodal documents are best conducted in multidisciplinary teams that combine linguistic expertise with domain-specific knowledge and competence in methods drawn from communication studies and computer science. It is this emerging field of intersection that we draw from and that we attempt to foster with this contribution.

We proceed in the following steps. First, we distinguish three approaches to identifying bias in media offerings and opt for a media-internal comparative strategy by which different language versions are contrasted to detect bias. Second, the specific features of multimodal Wikipedia entries are introduced alongside related information sources such as Wikimedia Commons metadata, changelogs representing the entries' version histories, and user discussions about such changes. Afterward, three analytical procedures are presented, which gauge different dimensions of bias between language versions:

1. an automatic text analysis of the political ideologies mentioned in entries on the same topic (i.e., Greek prime minister Alexis Tsipras and his party Syriza) using different language versions;
2. an automated sentiment analysis using the program Linguistic Inquiry and Word Count (LIWC) to compare the ratio of positively and negatively connoted words across language versions; and finally,

3. a manual analysis of the selection and textual embedding of images used in different language versions of the same topical entries, based on Wikimedia Commons metadata and changelogs.

These exploratory analyses are presented with the aim to develop a set of procedures that allow researchers to effectively and efficiently identify bias in Wikipedia entries as well as other factually oriented multimodal documents, particularly print and digital news offerings.

2 Background

The history of media and communication research abounds with studies that have tried to measure media bias or neutrality. Often such studies were conducted under the label of investigating "objectivity" because in Western journalism the idea of "objective reporting" became the dominant ideal in the period after the Second World War, complementing or even sidelining traditional ideals of interpretative or advocacy journalism in Western countries (Esser and Umbricht 2014). That trend in normative thinking may actually be reversing with the enormous expansion of the "space of opinion" (Jacobs and Townsley 2011) in legacy media and the World Wide Web as well as the continuous development of longer form, more interpretive newspaper journalism even in the United States (Barnhurst and Mutz 1997). But even if objectivity remains a powerful idea, the problem in the context of this chapter is that it is even harder to pinpoint than neutrality or bias. Neutrality can be considered a sub-dimension of objectivity, and an element that—especially in contrast to its opposite, bias—explicitly evokes measurable notions of evenhandedness and balance. But more questions remain: Which universe of entities or elements should be presented in a balanced fashion to achieve neutrality, and in which features of multimodal presentation should that balance be manifested? And where do the standards originate against which actual performance levels should be evaluated?

In principle, there are three possibilities to answer the latter question, all of which have been dealt with in media and communication research extensively (Donsbach 1990):

1. Extra-media comparisons. In this strategy, media accounts are compared to external sources of knowledge such as direct human observation of an event, statistical indicators on real-world phenomena, or expert opinion.

However, reliable extra-media information is often difficult to find, and its production often relies on institutional logics not directly comparable to media production, so that the extra-media comparison does not help to validly assess the neutrality of the media account.
2. Intra-media comparison. Here the accounts of different types of media or media offerings are compared to gauge *relative* levels of bias and neutrality. By identifying individual offerings that deviate from an empirical mainstream in presenting an issue or a person, or by identifying two or more distinct variants of presentation, this strategy can single out more or less biased or neutral versions without recourse to an absolute external standard.
3. Measures of equal distribution. This third strategy presupposes that a fixed set of strictly opposing attributes can be identified (such as critical or affirmative comments, positively or negatively connoted adjectives, left-leaning or right-leaning ideologies, etc.) and posits that both poles of each attribute should be presented with the same frequency. Even though Option 3 is the closest to the literal meaning of balance, its use is limited to very specific cases in which (1) such fixed opposing attribute sets exist, and (2) equal distribution is actually a plausible goal. In most real-world media accounts neutrality is not achieved through strict equality of polar attributes but through more variable and multifarious forms of textual and visual realization.

The exploratory work presented in this chapter relies on Option 2. By comparing different language versions of the same Wikipedia entries, differences in describing and depicting an object are unearthed. The size of such differences hints at the leeway that Wikipedia authors in different language environments use to provide distinct accounts. In this chapter bias is, thus, not assessed in relation to some absolute external standard, but it resides in the relative deviation of language versions from each other. The distribution of such deviations—in terms of outliers versus mainstream groups of language versions in relation to a particular set of features—reveals the overarching patterns of bias. The result is an entirely media-internal measure of relative bias and neutrality.

Wikipedia is one of the most successful projects of the social web: Since its launch in 2001, thousands of contributors have built this huge knowledge resource, which is not only used as an online encyclopedia but also as an object of research in various academic disciplines. Wikipedia is a multilingual resource that is currently available in 280 active language versions (cf. List of Wikipedias 2016), whose articles are connected via inter-language links. Users can activate these links to switch between articles on the same topic provided in different languages. Comparative studies, like Hecht and Gergle (2010), Franz (2011), or Young-Ho et al. (2015), found that these articles are in most cases not simple translations of each other but differ considerably in size and content. Wikipedia is, thus, an interesting object for cross-language and cross-cultural research.

Wikipedia articles comprise images and other kinds of media objects (e.g., audio and video files), they can thus be regarded as multimodal documents (Bateman 2008). Figure 9.1 shows an extract from the article on "Alexis Tsipras" in the English Wikipedia. This snapshot was taken in September 2015 to illustrate the main building blocks for our analysis. Most images in Wikipedia articles are accompanied by a caption. Images and their captions form clusters that are inserted in the text body of the article in different positions: the first image in Figure 9.1 is displayed on the left side, while the second and third images are positioned on the right side of the article. In contrast to textbooks or scientific articles, the image-caption clusters in Wikipedia are not numbered. It is, thus, impossible to explicitly refer to these clusters by numbers. However, in many cases lexical cohesion (in the sense of Halliday and Hasan 1976) ties the image-caption cluster to the surrounding text body. An example is the second image-caption cluster displayed in Figure 9.1: the segments "laying down red roses" and "Memorial" occur in both the caption and the text body. We use the term "cohesion anchor" for elements in the text body that contain lexical overlap with the caption text, and the term "anchoring text segment" to refer to the sentences that contain these anchors. In our study we focus on the cross-language comparison of captions and anchoring text segments related to the same image in different language versions. We did not go into a deeper analysis of coherence and cohesion in these Wikipedia articles.[1] The central task for our purpose was to localize anchoring segments in the text bodies. It has to be noted that not all image-caption clusters in Wikipedia articles have such anchoring segments. The first image on the upper left side in Figure 9.1 is a non-anchored example: The caption describes the pictured event and provides context information, but neither the event—Tsipras' speech in Bologna—nor the Other Europe allied party mentioned in the caption is explicitly treated in the text body.

Most of the media objects embedded in Wikipedia entries are stored in Wikimedia Commons, a repository for media files that is also built as a collaborative project and supported by the Wikimedia Foundation (cf. Viégas 2007; Hammwöhner 2013). The media objects of Wikimedia Commons can be directly embedded in the entries without the need to separately upload them again. Wikimedia Commons provides detailed information on each of its media

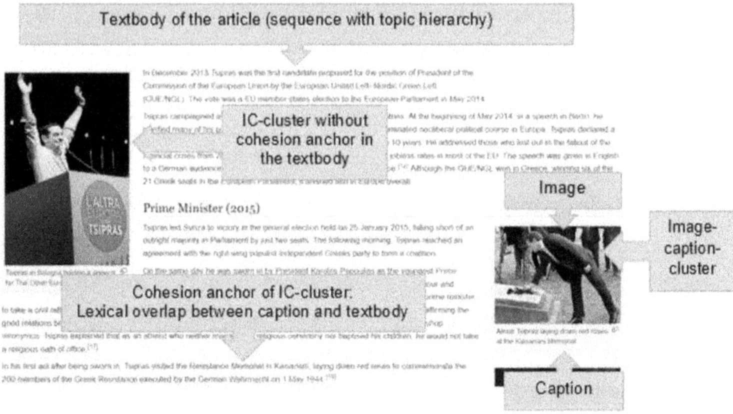

Figure 9.1 Extract from the article on "Alexis Tsipras" on the English Wikipedia (September 2015).

objects, inter alia information on the author, date, source, and licensing. It also contains a verbal description of the media object, which was used in our exploratory study as a "context-neutral" starting point for comparing caption texts in different language versions (see below). The descriptions of the media objects also comprise links to all Wikimedia projects in which the file is currently included. Among these projects are all language versions of Wikipedia and Wiktionary. Comparative studies on the use of images and other media objects can build on these link collections.

There is a growing amount of research on Wikipedia worldwide, and a variety of tools are already available for computational analysis. Wikipedia has been recognized as a valuable resource providing lexical semantic information for natural language programming (NLP) applications. Zesch (2009: 15) gives an overview about how different types of lexical semantic information can be exploited in the context of Wikipedia structure, content, and usage mining. Studies on the multilingual Wikipedia can build on tools like Omnipedia (Hecht and Gergle 2010; Bao et al. 2012) and Manypedia (Massa and Scrinzi 2013)[2], which offer various methods for cross-language research. A tool specialized in cross-language image analysis is offered by the Digital Methods Initiative[3]. The tool takes a Wikipedia article as a seed to select the articles of other language versions that are related by inter-language links and generates a table with all images in these articles. The table represents the order in which the images occur in their articles, but the image captions are stripped off. Rogers (2013: 191–200) demonstrates how this tool can be used as a component of a comprehensive method on cross-cultural Wikipedia analysis, which is exemplified by Wikipedia articles on the "Srebrenica massacre" in their English, Dutch, Bosnian, Croatian, Serbian, and Serbo-Croatian language versions.

From the very beginning, Wikipedia has also been conceived as a resource to study collaborative writing processes (cf. Ferschke, Daxenberger, and Gurevych 2013). In the first years of Wikipedia peer-production, the possible impact of hidden agendas and propaganda has been seen as a risk that might compromise the neutral nature of encyclopedic knowledge (Denning et al. 2005). In the meantime, sophisticated mechanisms have been introduced to prevent misuse of the editing functions and guarantee neutrality as far as possible (Kittur et al. 2007), and different forms of "systemic bias" are discussed within the Wikipedia community[4]. Despite these efforts, recent studies have shown that bias is still an important issue (e.g., Reagle and Rhue 2011). Greenstein and Zhu (2014) have investigated whether there is political bias toward Democratic or Republican positions in encyclopedias using a terminology-based analysis that has been used to detect political bias in mass media. The study shows that, indeed, the English version of Wikipedia is likely to be more biased toward a Democratic point of view than the Encyclopædia Britannica. The case study in Roessing and Podschuweit (2011) suggests that this bias has an impact on electoral campaigns. Our exploratory study is, therefore, concerned with the question of whether and how Wikipedia is used as a medium for political influence.

3 Studying bias and neutrality in Wikipedia cross-culturally: Party competition and the Euro crisis

In order to answer the research question "Is Wikipedia content really neutral or does it reflect the position of the author or the social context?" this study focuses on one prominent issue that has been extensively debated in all countries of the European Union, in particular in those that belong to the Eurozone, since the end of 2009. Because of the globalization of the economy and of the financial markets, and due to the international financial and banking crisis in 2007–08 and the recession of the following years, five member states of the Eurozone—Cyprus, Greece, Ireland, Portugal, and Spain—were not able to repay their government debt or to bail out over-indebted banks under their national supervision without the assistance of other Eurozone countries, the European Central Bank (ECB), or the International Monetary Fund (IMF). Because of the austerity policy implemented in these states, a significant share of the population experienced a worsening of living conditions. The national governments, regardless of their ideological background, had to implement severe cuts of welfare expenditures and to increase tax rates. One reaction was a significant decrease of trust in political institutions, political parties, and in democracy in general (e.g., Armingeon and Guthman 2014). This resulted in strong opposition against the parties that formed the incumbent governments: citizens in those states organized demonstrations, created new political parties, or started supporting parties that hitherto had only received small vote shares in national elections.

While the Spanish party Podemos ("we can") is an example of a newly established and electorally successful party in the aftermath of the European debt crisis, the Greek left-socialist party Syriza can be seen as an example of an already established political party that gained enormous support in the course of the Euro crisis. While Syriza received between 3.3 and 5 percent of the vote in the three Greek national elections in the time period between 2004 and 2009 and was considered a small, far-left "niche party" (see, e.g., Rovny 2013; March and Rommerskirchen 2015), its vote share increased significantly in the face of the crisis, so that Syriza became the major opposition party after the 2012 parliamentary election and the strongest government party in 2015. Syriza party leader Alexis Tsipras assumed office as Greek prime minister in January 2015 and acquired widespread media visibility for his opposition to the austerity measures pushed forward by the European Commission and several EU member state governments, in particular the German government led by Chancellor Angela Merkel.

Alexis Tsipras, members of his coalition cabinet—in particular Finance Minister Yanis Varoufakis—as well as their economic and financial policy were highly controversial among the European public and got more favorably evaluated in countries that were also heavily affected by the European debt crisis (and more critically in states that were not hit strongly by the crisis). Therefore, the study presented here starts with the expectation that Syriza and Alexis Tsipras are presented differently across Wikipedia language versions, both in terms of how the ideological position of Syriza is identified and the way in which Tsipras is presented personally. The key explanatory variable is the role of the country in the Euro crisis. As the German government adopted a very rigid pro-austerity policy, we hypothesize that Alexis Tsipras is presented more negatively and Syriza more radically in the respective German language Wikipedia entries, whereas Tsipras should be more positively framed and Syriza identified as more moderate in the Spanish Wikipedia entries. This is because Spain was also hit strongly by the economic crisis and a party similar to Syriza—Podemos—has been very successful in Spanish elections since 2014. By contrast, the Wikipedia entry in the English and French versions should display rather neutral positions. Although both countries mentioned last were also affected by the financial and economic crisis and see an increasing vote share of far right-wing parties like the Front National in France and the United Kingdom Independence Party, Spain and Greece received special support from international organizations and EU member states to solve their problems. So we expect to find a stronger biasing effect in the case of the Spanish, Greek, and German language versions of the Wikipedia entries about Tsipras and Syriza.

3.1 Using ideology descriptions to identify bias

In order to identify how Syriza is described in the Wikipedia entries across languages, we refer to one common feature of Wikipedia entries on political parties: the mentioning of an ideology in the short description of the party's profile provided in a separate box on the right-hand margin of the entry. For instance, the entry for the British Conservative Party on the English Wikipedia mentions four ideologies that help characterize succinctly the programmatic profile of the Tories: Conservatism, economic liberalism, British unionism, and soft Euroscepticism. In case of the British Labour Party, by contrast, the English Wikipedia mentions two ideologies—social democracy and democratic socialism—to describe the overall programmatic position of the current main opposition party in the House of Commons. Such basic ideological orientations are often used in political science to describe party systems, patterns of party competition, office allocation within coalition cabinets, or legislative speechmaking (e.g., Budge and Keman 1990; Bäck et al. 2011; Bäck and Debus 2016). For instance, relying on studies about party families in Western Europe and historical studies of social cleavages and groups traditionally supporting particular parties as well as on analyses of supporters' attitudes and of parties' programmatic statements, Budge and Keman (1990: 95) distilled a ranking of portfolio types for five central party families—Conservatives, Liberals, Religious, Socialists, and Agrarians—and assumed that the preferences of parties belonging to one of these ideological families are very similar. The recent literature in party research differentiates between eight party families: ecological, (former) communist, social democratic, liberal, Christian democratic, conservative, agrarian, and nationalist (see the coding scheme of the Comparative Manifesto Project, provided in Volkens et al. 2013). However, party competition in Europe became more fluid in the last decades, so that a number of new parties with a new ideological background emerged. These include, for instance, Eurosceptic parties, antiglobalization parties, or regionalist parties (for an overview, see Caramani 2014). These new "ideologies" are described in detail in separate Wikipedia articles and are mentioned in the Wikipedia entries on those political parties that represent the respective ideologies according to the authors of the Wikipedia entries.

The goal is to estimate the "ideological position" of the ideologies mentioned in the description of Syriza in the Wikipedia entries of the five languages: English, French, German, Greek, and Spanish. A fully computerized method of content analysis—Wordfish (Slapin and Proksch 2008)—was applied to the full text of the English Wikipedia entries of the respective ideologies. Fully computer-aided methods of content analysis like Wordscores developed by Laver, Benoit, and Garry (2003) and Wordfish by Slapin and Proksch (2008) are advancements of semi-manual "dictionary approaches" (for an overview, see Debus 2009). The main advantage of both approaches is that the position estimation is left completely to computer algorithms. Therefore, potential problems associated with purely manual procedures relying on dictionaries do not arise. The basic idea of both the techniques is to compare the frequency distribution of words from different texts and to make conclusions concerning the position of a text for specific policy areas on the basis of differences in the share of words used within the set of analyzed political documents.

Figure 9.2 shows the estimated positions of the ideologies mentioned on the Wikipedia entries of selected English, Greek, and Spanish parties.[5] The ideological dimension—extracted on the basis of the words mentioned in the Wikipedia entries of the respective ideologies—nicely reflects a general left-right continuum. While ideologies like communism or democratic socialism receive negative Wordfish scores, ideologies that refer to nationalism are located on the opposite side of the dimension. Those ideologies that appear in the Greek version of the Syriza Wikipedia entry are highlighted in Figure 9.2. Save the entry on secularism, all remaining ideologies—eco-socialism, libertarian socialism, Eurocommunism, democratic socialism, and communism—are located on the far-left side of the ideological scale.

In the following step, the arithmetic mean of the scores on the six ideological dimensions with which the Greek Wikipedia entry describes the ideological background of Syriza is used to estimate the ideological position of this political party in the Greek Wikipedia. We apply the same procedure for the other languages under study—English, French, German, and Spanish. Figure 9.3

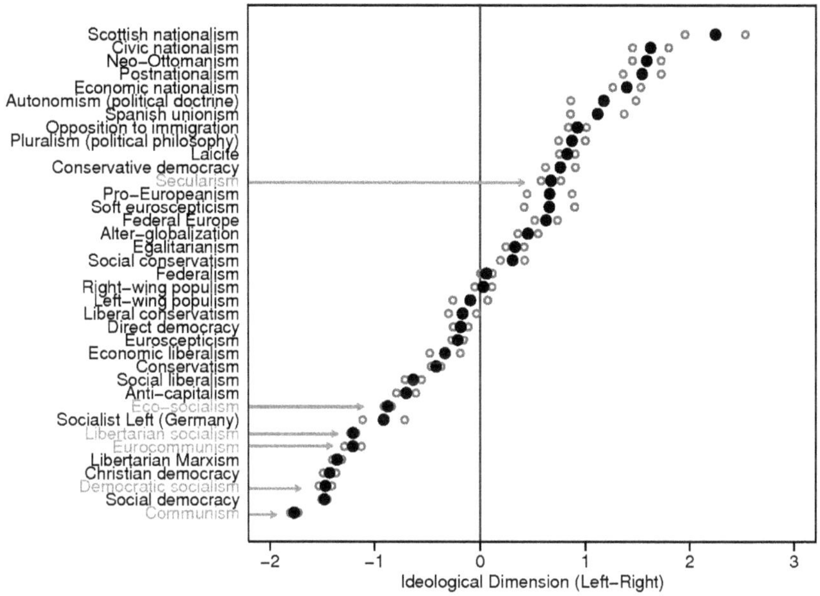

Figure 9.2 Estimated positions of the ideologies mentioned in the Wikipedia entries on selected English, Greek, and Spanish parties.

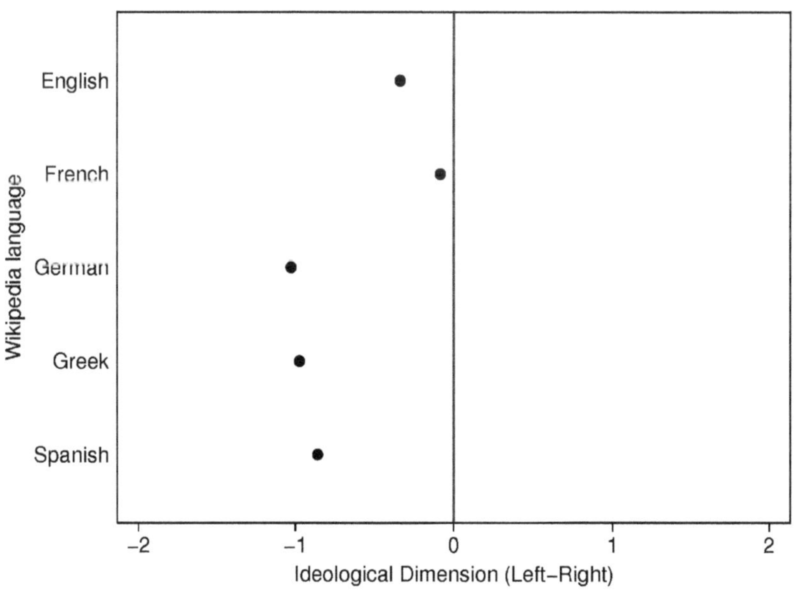

Figure 9.3 Mean ideological scores for "Syriza" in different Wikipedia language versions.

shows the mean ideological score for the Syriza entry in the respective language versions. The findings indicate that, in contrast to the expectations, the position of Syriza is more extreme and to the far-left of the ideological dimension not only in the German but also in the Greek and Spanish versions of the Syriza Wikipedia entry. In addition, Syriza is described by more "moderate" ideologies in the French and English entries. However, we find clear empirical evidence that the content of a Wikipedia article—at least in terms of the ideologies mentioned in the Syriza Wikipedia entries—does vary across language versions. A reader of the German Wikipedia entry on Syriza is likely to perceive the party as ideologically extreme on the left-right dimension, whereas a reader of the English or French article version should consider this party rather moderate in ideological terms.

3.2 Sentiment analysis with LIWC

As stated above, we expect that Alexis Tsipras will be framed rather negatively in the German version of Wikipedia, neutrally in the English and French, and positively in the Spanish and Greek versions. To assess this hypothesis, we carried out a sentiment analysis of Tsipras's Wikipedia articles in all existing language versions as of September 12, 2015. Doing so, we used LIWC, which is able to evaluate the sentiment of texts across a set of over eighty categories (Tausczik and Pennebaker 2010: 27). Using LIWC2015 (Pennebaker Conglomerates Inc. 2015), we solely focused on its category "Affect Words" or, more specifically, the subcategories "Positive Emotion" and "Negative Emotion." Based on its internal dictionaries assigning each contained word an emotional score, LIWC was used to calculate the percentage of positively and negatively connoted words in Tsipras' article for every language.

As LIWC dictionaries only exist for a limited set of languages, all non-English versions of Tsipras' article were first translated into English. This step was performed automatically by using the Python library TextBlob (Loria 2015). While such an automated translation might contain both semantic and syntactic flaws, we argue that the latter can be neglected in our process. This is because LIWC assigns its scores based on a word-level, not on a sentence-level comparison with the dictionaries. Hence, possible grammatical errors caused by the machine translation do not affect the results. Out of the sixty-seven distinct language versions, our process was able to create sixty-one automated translations—the six languages that could not be translated were Bavarian, Upper Sorbian, Indonesian, Lombard, Scots, and Simple English.

However, some of the automated translations missed large parts of the original article and thus had to be omitted. Therefore, a simple sanity check was performed as follows: LIWC provides its users with both the total word count within an analyzed document and the number of a document's words that were successfully recognized in the program's dictionaries. In the case of Tsipras's article, with a mean word count of ~506 and ~59 percent of the words identifiable by the dictionaries on average, the product of both numbers (~506 * ~.59 = ~299) denotes the mean number of recognized words across all language versions. To ensure a critical length, all language versions with less than 299 recognized words were disregarded in the further process. This sorting yielded a final sample of eighteen languages with an above-average number of words recognized by LIWC.

To gather insights about the polarity of these remaining eighteen language versions, we compared their "Positive Emotion" and "Negative Emotion" scores. Research shows that, in general, LIWC's ratings for these two categories match human assessments (Alpers et al. 2005). Hence, we are confident that the result of our sentiment analysis gives an approximation of the bias contained in each language version. While the Hebrew, Slovak, and Dutch versions of Tsipras' article yielded the highest share of emotionally connoted words (5.72, 5.65, and 4.84 percent, respectively), the Italian, Georgian, and Turkish have the lowest shares (2.48, 2.38, and 1.94 percent). The mean sum across all languages is 3.43 percent. To set both positive and negative emotional scores in relation, a ratio was calculated for every language version:

$$\textbf{\textit{Emotional Score Ratio}} = \frac{\textbf{\textit{Positive Emotional Score}}}{\textbf{\textit{Negative Emotional Score}}}$$

Figure 9.4 shows the results of our analysis with all language versions depicted in decreasing order according to their emotional score ratios.

With a mean ratio of 3.15, it can be inferred that, on average, all language versions feature three times as many positively connoted words as negative ones. Furthermore, with 9.78, the Turkish version of the entry possesses the highest emotional score ratio, hinting to an exceptionally strong positive connotation of the "Tsipras" article. With a large gap, the Georgian (4.95) and Polish (4.23) versions follow. In contrast, the Dutch version with a ratio of 1.15 indicates an almost equal distribution of positive and negative scores. The Japanese and the Slovak versions have only marginally larger ratios (1.47 and 1.50). Since there is no language with an emotional score ratio below 1, it can be inferred that no article is more negatively than positively connoted.

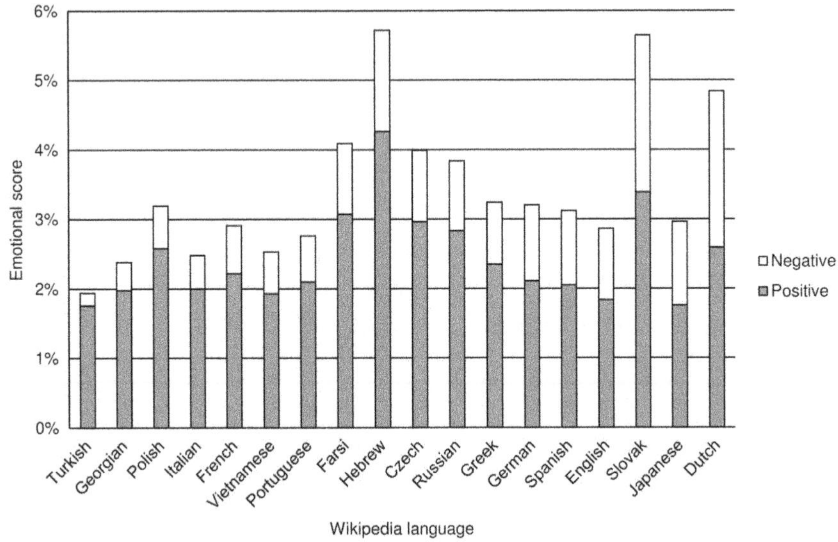

Figure 9.4 LIWC Sentiment estimates for different language versions of the "Tsipras" entry.

Contrary to our expectation that Tsipras would be displayed rather negatively in German, neutrally in English and French, and positively in Spanish and Greek, the French version is rather positively connoted (3.22). Moreover, the Greek, German, Spanish, and English versions all rank relatively low (2.64, 1.94, 1.92, and 1.80, respectively), which means they are all weakly positively connoted. Even though our initial hypothesis could not be confirmed, this sentiment analysis speaks to the more general question of bias: As all analyzed language versions are at least weakly positively connoted, all of them fulfill the criterion of being biased in the sense that they are unbalanced in their use of adjectives, with Turkish being the most and Dutch being the least biased language version. To investigate this outcome further, a more in-depth theoretical framework should be developed, and the process of the empirical analysis should be further refined. As LIWC is a program mainly intended for research in psychology (Tausczik and Pennebaker 2010) instead of political science or communication studies, dictionaries specifically suited for investigating political communication should be established in the future.

3.3 Manually comparing image selection and use

A core question with respect to the multimodal focus of this work is whether differences in the description of the same topic or entity are only reflected in the language used and the references made or also in the images presented in the respective articles. This concerns two aspects:

1. Image selection: Do articles in different language versions use different images to illustrate the same topic or entity, and, if so, do the different images take different viewpoints on the subject matter?
2. Image contextualization: If the same image is used in two or more different language versions, are there differences in the way the articles contextualize the image to highlight certain aspects?

Concerning the contextualization of images, the image caption and at the anchoring text segment in the article were used. The latter is not always available, as images in Wikipedia are not placed inside the text and there is no obvious way of referring to them because image numbers are lacking. Nevertheless, article texts often refer to the motif of accompanying images and thus anchor the image in the text. Entries on different European parties and politicians associated with the financial crisis were analyzed with respect to the two aspects mentioned. No general conclusions about the systematic use of images can be drawn so far, but some interesting differences were observed. In the following, we return to the Wikipedia entries about the Greek prime minister Alexis Tsipras.

Looking at the portraits of Tsipras used in the language versions, three different role conceptions can be clearly distinguished.

Figure 9.5 Different role attributions in portraits of Alexis Tsipras.

Figure 9.5 shows examples of these three different roles from the Spanish, the German, and the Greek versions of Wikipedia: "the person," "the politician," and "the statesman," respectively. The statesman role is represented by photos taken during the swearing-in as prime minister. Images attributing the politician role show Tsipras giving speeches, typically against a red background, Syriza's party color. Images that show Tsipras in front of a neutral background without any hint to his profession or position are labeled "the person" view.

The rationale for selecting a particular role attribution is not always obvious. We cannot explain, for example, why the German page uses the politician view while the Spanish uses the person view. We cannot rule out the possibility that in some cases the choice might be pragmatic in nature because the author happened to come across one or the other image first. However, there is a strong indication that image choice is often intentional, particularly when the Wikipedia changelog shows frequent replacements of an image by different people. An example can be found in the English Wikipedia where the image of Tsipras was replaced several times within a short time period. Before July 27, 2015 the page showed an image taking the "politician" view. On that day, it was changed to the "statesman" view, and on the next day (July 28, 2015) it was changed back to the previous depiction. We can only speculate that this change between depicting Tsipras as a political contender for power and as the official power holder is related to different Wikipedians' views on his legitimacy and achievements. On August 20, the day Tsipras announced the resignation of his first cabinet, the image was replaced by one taking the "person" view again, possibly signifying the loss of power and the necessity to forge a new power base. Such changes between different versions of the same motif suggest that the choice of a particular depiction is indeed a strategic issue.

Concerning the second aspect, that is, the contextualization of one and the same image in different language versions, the image shown in Figure 9.6 can serve as an example. It appears in many different language versions and shows Alexis Tsipras during his first official act as prime minister. The master version of this image, which was uploaded to Wikimedia Commons in March 2015, carries the description, "Greek Prime Minister Alexis Tsipras lays down red roses at the National Resistance Memorial, Kaisariani on 26 January 2015." A comparison of this description with the image captions and anchoring text segments used in different language versions is shown in Table 9.1.

Figure 9.6 Image of Alexis Tsipras during his first act as prime minister.

Some aspects of the image content are reported in all versions in more or less the same way. In particular, almost all language versions analyzed here mention that the image shows the first official act of Alexis Tsipras as prime minister, often both in the caption and the anchoring text. This uniformity can be explained by the symbolic character of laying roses at the Kaisariani memorial. Differences in how the nature of the act was perceived can be seen in the references to the victims to whom Tsipras is paying his respect. In the Wikimedia description, they are not mentioned at all. The Greek version only provides a very neutral reference, speaking of "those executed" without any further details. The German version is a bit more specific and refers to them as "victims of the German Wehrmacht," though still without further background. The English version is more specific and refers to "200 members of the Greek resistance executed by the German Wehrmacht." The Spanish version, finally, adds the information that the victims were of the "Greek Communist resistance," but it does not directly attribute the killing to the German Wehrmacht, rather mentioning "the occupying forces of the Axis." This example shows that there are notable differences in the choice and the contextualization of images in different Wikipedia language versions. While there are strong indications that these differences are not arbitrary but probably reflect the perspective of the authors, a detailed analysis of that perspective is impossible on the basis of the Wikipedia entries alone. And this is even more difficult in multimodal documents meant to avoid perspective and bias such as encyclopedia entries, which do not feature clear and unequivocal cues on their creators.

Table 9.1 Image context in different Wikipedia language versions.

Wiki resource	Image caption	Anchoring text segment
Wikimedia Commons	Greek prime minister Alexis Tsipras lays down red roses at the National Resistance Memorial, Kaisariani on January 26, 2015.	
English Wikipedia	Alexis Tsipras laying down red roses at the Kaisariani Memorial.	In his first act after being sworn in, Tsipras visited the Resistance Memorial in Kaisariani, laying down red roses to commemorate the two hundred members of the Greek Resistance executed by the German Wehrmacht on May 1, 1944.
German Wikipedia	First official act of Tsipras—commemoration of the victims of the German Wehrmacht in Kaisariani.	As first official act, Tsipras lays down flowers at the memorial on the shooting range of Kaisariani
Spanish Wikipedia	Tsipras's first act as prime minister was to visit the National Memorial of Resistance at Kaisariani where two hundred members of the Greek communist resistance were shot in 1944 by the occupying forces of the Axis.	--
Greek Wikipedia	Tsipras leaves roses on the Kaisariani shooting range after being sworn in as prime minister.	The first move after the inauguration was the tribute to those executed at the shooting range of Kaisariani

Therefore, the creation process is not yet understood well enough to judge whether the differences between language versions observed here are indicators of different political positions that surface despite Wikipedia's strong emphasis on neutrality. Luckily, Wikipedia is different from other media products in that it opens a window into the creation process itself by offering changelogs and explicit discussion about editorial choices for users and analysts to observe.

In order to get a more complete picture of the role that images and captions play in political influence, more types of information coming from other Wikipedia namespaces could be examined, for example, discussions of image captions on talk pages and information on the users involved in the editing processes. As Wikipedia authors should be aware that they are publishing for the general public, analyzing the information they give away about themselves does not constitute a particular ethical issue. Researchers could thus take a closer look at the processes of image-related editing and the policies and informal rules for the use of images within the community. Even more in-depth information could come from conducting interviews with users involved in editing as well as those experienced Wikipedians who oversee compliance to Wikipedia policies.

4 Conclusion

The exploratory work presented here serves to demonstrate that it is worthwhile to integrate the multimodal perspective into cross-lingual Wikipedia analysis. The selection of images and their contextualization via captions and anchoring text segments constitute distinct forms through which bias is created in Wikipedia entries. They complement the lexical choices that authors of different language versions make in using emotionally charged adjectives and more abstract conceptual descriptors. The effect of such choices on both the textual and visual content as well as on their relation in concrete multimodal compositions can be nicely elucidated through media-internal comparison, that is, the comparison of different language versions of one and the same entry as presented in this chapter. Such internal comparison is better suited to detect bias and neutrality than its alternatives: it is almost impossible, on the one hand, to theoretically justify absolute external standards of comparison against which individual items or elements can be compared. On the other hand, it is rarely appropriate to use standards of equal distribution for discourse elements in factual accounts that were collaboratively produced by nonexperts. Having said that, there is a lot of room to further refine, validate, and extend the work presented in this chapter.

As a first step, it is paramount to ascertain whether lexical choices as identified by sentiment analysis and image choices as determined by visual comparison covary in meaningful ways. A follow-up study should thus combine and relate the analytical procedures described above so that the multimodal character of bias creation becomes transparent. Furthermore, the methods sketched here should be evaluated on a broader empirical basis by systematically checking whether levels and forms of bias creation are dependent on the issues covered. Some issues may resonate with convictions held by particular cultures or language communities more strongly than others—and may thus exhibit a higher level of bias. Moreover, automated and manual methods should be combined in a more systematic fashion to create a mixed-methods toolkit. As part of this endeavor, the computational methods could be adapted to the specific task of cross-language and multimodal Wikipedia research. For instance, sentiment analysis could be improved if the LIWC dictionaries were customized to the vocabulary relevant for political communication. Textual sentiment analysis could also be complemented with visual methods that automatically detect emotions through facial features (Bettadapura 2012; Samal and Iyengar 1992). Conversely, the manual analysis of image selection and contextualization could be supported by more elaborate (semi)automatic methods that compare caption texts assigned to the same image in different language versions. Tools to identify and visualize image-related edits and reverts in article histories would help investigate the processes of image and caption editing in a more systematic fashion. Ultimately, in those cases where other media objects (especially video files) are embedded in an entry, these more complex multimodal objects should be included in the analysis.

One final issue concerns the question of whether bias and neutrality in Wikipedia entries should be compared to other types of media offerings. Given that Wikipedia is explicitly geared toward providing neutral, factual accounts, it would be most instructive to see how Wikipedia entries fare in comparison to, for example, quality journalism inspired by the ideal of objective reporting. Professional news websites or quality print newspapers would, thus, seem to be ideal test cases. While it seems fairly clear that political content on other social media platforms such as Twitter debates and political discussions on Facebook will be heavily biased by political positions, in the case of professional news journalism the outcome seems less clear. Therefore, a large and interdisciplinary field for intra- and intermedia comparison lies ahead of researchers interested in analyzing bias and neutrality in multimodal settings.

Notes

1 Cohesion and coherence in non-sequential hypertext and in multimodal documents is discussed in Fritz 1999, Storrer 2001, Bateman 2014.

2 http://www.manypedia.com/.

3 https://tools.digitalmethods.net/beta/wikipediaCrosslingualImageAnalysis/.

4 See for example: https://en.wikipedia.org/wiki/Wikipedia:Systemic_bias.
5 These parties include Podemos (new left-wing populist party in Spain), the Spanish PSOE (socialist party), the Spanish PP (conservative party), the Spanish Ciudadanos (a liberal, centrist protest party), UKIP (English nationalist party), the SNP (Scottish nationalist party), and SYRIZA.

References

Alpers, G. W., A. J. Winzelberg, C. Classen, H. Roberts, P. Dev, C. Koopman, and C. Barr Taylor (2005), "Evaluation of Computerized Text Analysis in an Internet Breast Cancer Support Group," *Computers in Human Behavior*, 21(2): 361–76.

Armingeon, K. and K. Guthmann (2014), "Democracy in Crisis? The Declining Support for National Democracy in European Countries, 2007–2011," *European Journal of Political Research*, 53(3): 423–42.

Bäck, H., M. Debus, and P. Dumont (2011), "Who Gets What in Coalition Governments? Predictors of Portfolio Allocation in Parliamentary Democracies," *European Journal of Political Research*, 50(4): 441–78.

Bäck, H. and M. Debus (2016), *Political Parties, Parliaments and Legislative Speechmaking*, Houndmills: Palgrave Macmillan.

Bao, P., B. Hecht, S. Carton, M. Quaderi, M. Horn, and D. Gergle (2012), "Omnipedia. Bridging the Wikipedia Language Gap," in E. Mynatt (ed.), *Proceedings of the SIGCHI Conference on Human Factors in Computing Systems 2012*, 1075–84, New York: ACM Press.

Barnhurst, K. and D. Mutz (1997), "American Journalism and the Decline in Event-Centered Reporting." *Journal of Communicatio*, 47: 27–53. DOI:10.1111/j.1460-2466.1997.tb02724.x.

Bateman, J. A. (2008), *Multimodality and Genre: A Foundation for the Systematic Analysis of Multimodal Documents*, Basingstoke/New York: Palgrave Macmillan.

Bateman, J. A. (2014): "Multimodal Coherence Research and its Applications," in H. Gruber and G. Redeker (eds.), *The Pragmatics of Discourse Coherence*, 145–77, Amsterdam/Philadelphia: John Benjamins.

Bettadapura, V. (2012), "Face Expression Recognition and Analysis: The State of the Art," arXiv:1203.6722.

Budge, I. and H. Keman (1990), *Parties and Democracy: Coalition Formation and Government Functioning in Twenty States*, Oxford: Oxford University Press.

Caramani, D. (2014), "Party Systems," in: D. Caramani (ed.), *Comparative Politics*, 216–36, Oxford: Oxford University Press.

Debus, M. (2009), "Analysing Party Politics in Germany with New Approaches for Estimating Policy Preferences of Political Actors," *German Politics*, 18(3): 281–300.

Debus, M., H. Stuckenschmidt, and H. Wessler (2015), "On the Use of Different Modalities in Political Communication: Evidence from German Election Manifestos," in J. Wildfeuer (ed.), *Building Bridges for Multimodal Research. International Perspectives on Theories and Practices of Multimodal Analysis*, 211–25, Bern/New York: Peter Lang.

Denning, P., J. Horning, D. Parnas, and L. Weinstein (2005), "Wikipedia Risks," *Communications of the ACM*. Online: http://www.csl.sri.com/users/neumann/insiderisks05.html#186 (accessed February 1, 2016).

Donsbach, W. (1990), "Objektivitätsmasse in der Publizistikwissenschaft," *Publizistik*, 35(1): 18–29.

Esser, F. and A. Umbricht (2014), "The Evolution of Objective and Interpretive Journalism in the Western Press: Comparing Six News Systems since the 1960s," *Journalism and Mass Communication Quarterly*, 91(2), 229–49.

Ferschke, O., J. Daxenberger, and I. Gurevych (2013), "A Survey of NLP Methods and Resources for Analyzing the Collaborative Writing Process in Wikipedia," in I. Gurevych and J. Kim (eds.), *The People's Web Meets NLP. Collaboratively Constructed Language Resources*, 285–310, Berlin/Heidelberg: Springer.

Franz, G. (2011), *Die vielen Wikipedias. Vielsprachigkeit als Zugang zu einer globalisierten Online-Welt*, Boizenburg: Verlag Werner Hülsbusch.

Fritz, G. (1999), "Coherence in Hypertext," in W. Bublitz, U. Lenk, and E. Ventola (eds.), *Coherence in Spoken and Written Discourse*, 221–32, Amsterdam/Philadelphia: John Benjamins.

Greenstein, S. and F. Zhu (2014), "Do Experts or Collective Intelligence Write with More Bias? Evidence from Encyclopædia Britannica and Wikipedia," Working Paper, Harvard Business School 15-023. Online: http://hbswk.hbs.edu/item/do-experts-or-collective-intelligence-write-with-more-bias-evidence-from-encyclopdia-britannica-and-wikipedia (accessed October 1, 2014).

Halliday, M. A. K. and R. Hasan (1976), *Cohesion in English*, London: Longman Group.

Hammwöhner, R. (2013), "Bilddiskurse in den Wikimedia Commons," in B. Frank-Job, A. Mehler, and T. Sutter (eds.), *Die Dynamik sozialer und sprachlicher Netzwerke. Konzepte, Methoden und empirische Untersuchungen an Beispielen des WWW*, 121–60, Wiesbaden: Springer VS.

Hecht, B. and D. Gergle (2010), "The Tower of Babel Meets Web 2.0: User-Generated Content and its Applications in a Multilingual Context," in *Proceedings of the SIGCHI Conference on Human Factors in Computing Systems 2010*, 291–300, New York: ACM Press.

Jacobs, R. N. and E. Townsley (2011), *The Space of Opinion. Media Intellectuals and the Public Sphere*, Oxford/New York: Oxford University Press.

Kittur, A., B. Suh, B. A. Pendleton, and E. H. Chi (2007), "He Says, She Says: Conflict and Coordination in Wikipedia," *Proceedings of the CHI 2007*. Online: http://www-users.cs.umn.edu/~echi/papers/2007-CHI/2007-Wikipedia-coordination-PARC-CHI2007.pdf (accessed August 15, 2016).

Laver, M., K. Benoit, and J. Garry (2003), "Extracting Policy Positions from Political Texts Using Words as Data," *American Political Science Review*, 97(2): 311–31.

List of Wikipedias (2016), *Meta Discussion about Wikimedia Projects*. Online: https://meta.wikimedia.org/w/index.php?title=List_of_Wikipediasandoldid=15285106 (accessed February 1, 2016).

Loria, S. (2015), *TextBlob: Simplified Text Processing (Release v0.11.0)*. Online: http://textblob.readthedocs.org (accessed September 12, 2015).

March, L. and C. Rommerskirchen (2015), "Out of Left Field? Explaining the Variable Electoral Success of European Radical Left Parties," *Party Politics*, 21(1): 40–53.

Martinec, R. and A. Salway (2005), "A System for Image-Text Relations in New (and Old) Media," *Visual Communication*, 4(3): 337–71.

Massa, P. and F. Scrinzi (2013), "Manypedia: Comparing Language Points of View of Wikipedia Communities," *First Monday*, 18(1). Online: http://firstmonday.org/ojs/index.php/fm/article/view/3939/3382 (Accessed: 01 February 2016).

Moriarty, S. E. and M. N. Popovich (1991), "Newsmagazine Visuals and the 1988 Presidential Election," *Journalism Quarterly*, 68(3): 371–80.

Pennebaker Conglomerates Inc. (2015), "Linguistic Inquiry and Word Count". Online: http://liwc.wpengine.com (accessed October 1, 2015).

Reagle, J. and L. Rhue (2011), "Gender Bias in Wikipedia and Britannica," *International Journal of Communication*, 5: 1138–58.

Roessing, T and N. Podschuweit (2011), "Wikipedia im Wahlkampf: Politiker, Journalisten und engagierte Wikipedianer," in E. Schweitzer and S. Albrecht (eds.), *Das Internet im Wahlkampf: Analysen zur Bundestagswahl 2009*, 297–314, Wiesbaden: Springer VS.

Rogers, R. (2013), *Digital Methods*, Cambridge, MA: MIT Press.

Rovny, J. (2013), "Where Do radical Right Parties Stand? Position Blurring in Multidimensional Competition," *European Political Science Review*, 5(1): 1–26.

Samal, A. and A. Iyengar (1992), "Automatic Recognition and Analysis of Human Faces and Facial Expressions: A Survey," *Pattern Recognition*, 25(1): 65–77.

Slapin, J. B. and S. Proksch (2008), "A Scaling Model for Estimating Time Series Party Positions from Texts," *American Journal of Political Science* 52(3): 705–22.

Storrer, A. (2002), "Coherence in Text and Hypertext," *Document Design*, 3(2): 156–68.

Tausczik, Y. R., and J. W. Pennebaker (2010), "The Psychological Meaning of Words: LIWC and Computerized Text Analysis Methods," *Journal of Language and Social Psychology*, 29(1): 24–54.

Verser, R. and R. H. Wicks (2006), "Managing Voter Impressions: The Use of Images on Presidential Candidate Web Sites during the 2000 Campaign," *Journal of Communication*, 56(1): 178–97.

Viégas, F. B. (2007), "The Visual Side of Wikipedia," in R. H. Sprague, Jr. (ed.), *Proceedings of the 40th Annual Hawaii International Conference on System Sciences*, 85–94, Washington, DC: IEEE Computer Society.

Volkens A., J. Bara, I. Budge, M. McDonald, and H. Klingemann (2013), *Mapping Policy Preferences from Texts: Statistical Solutions for Manifesto Analysts*, Oxford: Oxford University Press.

Wessler, H., A. Wozniak, L. Hofer and J. Lück (2016), "Global Multimodal News Frames on Climate Change. A Comparison of five Democracies around the World," *International Journal of Press/Politics*, 21(4): 423–45.

Young-Ho, E., P. Aragon, D. Laniado, A. Kaltenbrunner, S. Vigna, and D. L. Shepelyansky (2015), "Interactions of Cultures and Top People of Wikipedia from Ranking of 24 Language Editions," *PLoS One*, 10(3): e0114825.

Zesch, T. (2009), "Study of Semantic Relatedness of Words Using Collaboratively Constructed Semantic Resources," PhD diss., Technische Universität Darmstadt. Online: https://www.tk.informatik.tu-darmstadt.de/fileadmin/user_upload/Group_UKP/publikationen/2010/PhD_TorstenZesch_SemanticRelatedness_2010.pdf (accessed February 1, 2016).

10

Exploring Organizational Heritage Identity: The Multimodal Communication Strategies

Carmen Daniela Maier

1 Introduction

An increasing number of companies, organizations, and institutions today strive constantly to be perceived through a "heritage prism" (Balmer 2013: 291). Heritage identity implementation and communication are considered profitable by all sorts of companies, organizations, and institutions, no matter their size or configuration, as "the continuity with the past illustrates an effort to achieve trust and recognition" and the corporate roots can be interpreted as "a vision for the future" (Wiedmann et al. 2013: 202). The turn toward the past and its undeniable significance for a successful future has become a persistent communicative strategy in the corporate context because a credible heritage can not only signal authenticity and reliability, but it can also contribute to stakeholders' stronger emotional attachment. For commercial entities, the benefit of implementing and communicating their heritage identity is related to "enhanced monetary worth," while for not-for-profit institutions, corporate heritage identity implementation and communication could help in "meeting their organization's societal and or charitable objectives over successive generations" (Balmer 2013: 292).

This phenomenon of implementing and communicating a heritage identity has generated a lot of interest in the field of corporate communication scholarship. The heritage research area is, thus, continuously growing by focusing on various issues, from conceptualizing heritage in relation to corporate identities (Balmer 2011, 2013) and brands (Balmer 2001; Blombäck and Brunninge 2009, 2013; Hakala, Lätti and Sandberg 2011; Urde, Greyser and Balmer 2007) to exploring the heritage communication and implementation strategies of various types of companies, organizations, and institutions in different contexts (Balmer 2011; Balmer, Greyser and Urde 2006; Brunninge 2009; Burghausen and Balmer 2014). However, when reading the corporate heritage literature, it becomes obvious that there is a notable scarcity of empirical studies related to not-for-profit organizations. Therefore, after exploring the heritage identity communication strategies of a commercial entity (Maier and Andersen 2015), the author has changed the research focus to not-for-profit organizations in order to address the above-mentioned gap.

By employing an interdisciplinary approach that combines corporate heritage identity theory, discourse analysis, and multimodality, this study explores the communication strategies used on the website of the Thomas Coram Foundation, "the UK's first dedicated children's charity with a fascinating heritage that spans 275 years" (Our history 2016). According to the foundation's webpage, every year "Coram helps children and young people develop their skills and emotional health, finds adoptive parents and upholds children's rights, creating a change that lasts a lifetime" (Our impact 2016).

The present study aims to create an understanding of the discursive strategies that are employed by Coram in order to communicate their heritage identity on their website, and simultaneously to provide a multimodal model of analysis that could be replicated in other studies focused on heritage identity communication. In order to meet these overarching objectives, the following research questions have been formulated:

1. How is "Coram" constituted discursively in a hypermodal context as a reliable historical organization?
2. How are multimodal legitimations strategically employed to communicate the trustworthy heritage identity of Coram?

By addressing these questions, this exploratory qualitative case study seeks to make a contribution to heritage identity research by moving it into a multimodal frame of reference and focusing on a not-for-profit organization. The chapter is structured as follows: After presenting essential definitions and conceptual basics related to heritage identity communication and multimodal discourse on which this study builds, the next section clarifies the methodological framework, and then the main findings are discussed. The findings are also summarized in order to derive directions and recommendations for future research on these methodological grounds.

2 Corporate heritage identity

As already mentioned, scholars in corporate communication have observed the growing prominence of historical references in various corporate materials targeted at various stakeholders. According to one of the founding fathers of corporate heritage research, the attention organizations accord to their heritage identity is motivated by the fact that "heritage identities not only have but also give an identity" (Balmer 2011: 1382). According to Burghausen and Balmer,

> The concept of heritage identity refers to the particular traits of an organization that meaningfully link its past, present and by referring to some aspect of an organization's past that is still deemed by current stakeholders to be relevant for contemporary concerns and purposes but that is concurrently perceived as worth to be maintained, nurtured and passed on to future generations. (Burghausen and Balmer 2015: 23)

It is also highlighted that heritage identity is manifested in multiple time strata (Balmer 2011, 2013). As a result, corporate heritage communication is "the organization's perennial communication of its core corporate identity traits" (Balmer 2013: 318).

Urde, Greyser, and Balmer clarify the difference between history and heritage, highlighting that history "explores and explains what is often an opaque past; in contrast, heritage clarifies and makes the past relevant for contemporary contexts and purposes" (Urde, Greyer and Balmer 2007: 6). From this perspective, "heritage is super historic" because it is more than just a part of history (Balmer 2013: 303). Researchers interested in historical references in organizations have acknowledged heritage as a salient means of constructing the identity of an organization, highlighting also that "the corporate identity is continuously (re)formulated" because "the interpretations of history can be adjusted to suit contemporary agendas" (Blömbæk and Brunninge 2009: 404). Chreim shows that researchers have pointed out the issue of "adaptive instability" (Gioia 1998: 22) of an organizational identity and clarifies that "stability in organizations can be achieved by adopting intangible or abstract identity attributes that allow for a variety of applications as the environment changes" (Chreim 2005: 568). Christensen and Cheney's research work confirms the need for organizations to balance these two issues—stability and change—in order to come across as "stable yet responsive entities" (Christensen and Cheney 2000: 258). In light of these findings, it is necessary to keep in mind the dynamic character of the heritage construct, as "heritage is subject to change, transformation and reinterpretation" (Balmer 2013: 303).

When exploring the organizations' strategic use of heritage, Burghausen and Balmer (2014) have identified a series of strategies for the implementation of the heritage dimension of a corporate identity. The following strategies have been identified as being employed in various activities and presented in various documents by managers: narrating, visualizing, performing, and embodying. Through these strategies, the organization can link "the past to the present and a potential future in a meaningful way" (Burghausen and Balmer 2014: 2318). The narrating strategy implies the usage of historical references that "establish links between the past, present and future across different types of communication" (Burghausen and Balmer 2014: 2318). In order to make these links more explicit, the visualizing strategy is also employed by using visual data that can include old and new photographs and other illustrations in relation to the textual elements. Both strategies can be manifested across a wide range of materials issued by an organization, from press releases and magazines to websites. "The actualization of corporate heritage identity through traditions, rituals and customs" constitutes the performing strategy (Burghausen and Balmer 2014: 2318). The last strategy, embodying, actualizes the manifestation of a heritage identity in material form. It can be manifested through, for example, corporate architecture, objects and spaces, traditional crafts, and "the personal identities of individual managers" (Burghausen and Balmer 2014: 2318). Thus, "the transtemporal nature of a corporate heritage identity" (Burghausen and Balmer 2014: 2319) can be identified by exploring these four strategies and their interplay in corporate discourse.

According to their findings, these implementation strategies of a heritage identity "support the notion of a multi-modal and multi-sensory identity system" because, as other identity researchers have also shown, "manifestations of corporate identities are invariably multi-modal and multi-sensory in nature" (Burghausen and Balmer 2014: 2313).

3 Multimodal discourse strategies

The growing "multi-modal and multi-sensory foci" mentioned by Burghausen and Balmer (2014: 2313) in connection with the more general developments in the wider corporate communication field can be pointed out also in the realm of discourse research (Jones 2012, 2013; Kress 2010; Kress and van Leeuwen 2001; O'Halloran and Lim 2014; van Leeuwen 2000, 2007, 2008, 2009a, b). Specifically, van Leeuwen highlights the integration of several semiotic modes in the discursive recontextualization of various social practices. Although he focuses on the linguistic realizations of several categories of legitimations that can be added to elements of social practices when these elements are recontextualized in discourse, he points out that "other semiotic modes can also recontextualize social practices" (van Leeuwen 2008: 22), and "some forms of legitimations can also be expressed visually, or even musically" (van Leeuwen 2008: 119). The categories of legitimations that can appear in the recontextualization of social practices in discourse are:

- *Authorization*: "legitimation by reference to the authority of tradition, custom, law and/or persons in whom institutional authority of some kind is vested";
- *Moral evaluation:* "legitimation by (often very oblique) reference to value systems";
- *Rationalization:* "legitimation by reference to the goals and uses of institutionalized social action" (van Leeuwen 2008: 105–6).

In the context of multimodal discourse analysis, attention has been given both to the multimodal recontextualization of social practices in discourses and to the complex roles of "intersemiotic complementarity" in contributing to discursive coherence across several semiotic modes and media (Royce 2007: 99). Royce underscores that "the relationship is synergistic in nature, a concept which describes the ability of elements, in the act of combining, to produce a total effect that is greater than the sum of the individual elements or contributions" (Royce 2007: 103). Furthermore, according to Lemke (2002, 2005, 2009), when multimodal communication takes place in a hypertextual environment, connecting meaning units with the help of hyperlinks also constructs "a traversal which is more than the sum of its parts" (Lemke 2002: 319).

It is evident that hypermodal communication contributes to the enhancement of the temporal conflations of past, present, and future mentioned by Burghausen and Balmer (2014), not only because a hypertext "blurs the distinction between what is inside and what is outside a text" (Landow 2006: 118), but also because "it facilitates the blurring of content-wise boundaries" (Maier and Andersen 2015, 2017).

4 Research method

The methodological challenges associated with such systematic and comprehensive explorations of multimodal and hypermodal data lie primarily in their interdisciplinary framework. In order to reveal how Coram's heritage identity is shaped multimodally in the communicative environment of their website, the present analysis identifies specific multimodal discursive strategies in the six top-level pages and forty-seven lower-level pages of the website. Table 10.1 specifies each of the pages that have been analyzed. The only third-level pages that have been analyzed are those belonging to one of the second-level pages, "Our history," which is one of the three subpages of the first top-level page called "The difference we make." This choice is motivated by the fact that "Our history" page is most closely connected to the topic of this study and it provides the most diverse legitimations of the organization's heritage identity implementation strategies.

Table 10.1 The top-level pages and the different subpages of Coram's website

Coram's website			
Top-level pages	**First-level subpages**	**Second-level subpages**	**Third-level pages**
The difference we make	*Our history*	*Thomas Coram, The Foundling Hospital, Thomas Coram School and after, Our governors, friends and supporters, Our musical heritage, Researching Coram's archive*	*Thomas Coram in America, Thomas Coram on educating girls, Thomas Coram asking for link, One fantastic fact about Thomas Coram, One song from Thomas Coram's playlist, Foundlings at war, Ten fictional foundlings, Tokens of love, The end of institutional care and afterwards, Dickens, Handel, Hogarth, Handel and the Messiah fantastic facts, Thomas Coram playlist*
	Celebrating 275 years of helping children		
	Our impact		
How we do it	*Adoptions, Our creative therapies, Supporting practitioners who work with children, Supporting parents, Supporting young people, Health and drug education, Upholding children's rights, Getting young voices heard*		
Support what we do	*Ways you can give, Ways you can get involved, Challenge events, Corporate partnerships, Trusts*		
Sharing our experience	*How we influence practice and policy, Our pioneering projects, Our local authority partnerships, Coram-I improvement services, Resource library*		
News and events	*News, Press office, Social and community events, Adoption events, Training and professional development events*		

The analytical work, thus, focuses on exploring "the heritage footprints" (Balmer 2011: 1386) in the fifty-three screenshots of the respective webpages, and each screenshot has been coded by employing two types of selected *a priori* codes, which have been presented above: van Leeuwen's (2000, 2007, and 2008) legitimation types and Burghausen and Balmer's (2014) categories of heritage identity implementation strategies. The analysis has also been performed by taking into consideration the linking affordances of the hypertext in order to elaborate on how clusters of multimodal legitimations contribute to the consistent representation of the heritage identity across the pages of Coram's website, and for highlighting the effect of those affordances upon the involvement of the website's users.

Table 10.2 exemplifies how the embodying and performing strategies have been identified on each page through correlating the multimodal and hypermodal resources employed for expressing specific types of legitimations. *Legitimations* are the main discursive strategies that have been selected to show how "continuity and change are managed discursively" (Chreim 2005: 567) because they contribute most to the validation of the chosen heritage identity implementation strategies. The legitimations are explored not only in their multimodal but also in their hypermodal form, as a series of various links and their interplay contribute to enhancing their discursive power in relation to the representation of the implementation strategies.

Table 10.2 An example of the multimodal transcription and coding

Page name and level	**Multimodal representations of heritage identity implementation strategies**		
	Verbal resources	**Nonverbal resources**	**Strategy types Legitimation types**
Our history (top level) *Our governors, friends, supporters* (second level)	*The commitment of figures such as Handel, Hogarth, and Dickens helped establish a creative tradition at the Foundling Hospital that is continued today in the work of our Creative Therapies service.*	Three images: Hogarth's bust, Handel's portrait, and a black-and-white photo of Dickens Hyperlinks: each of these images	*Embodying strategy:* Verbal expert and role model authority legitimations accompanied by three hyperlinked images accompanying the text *Performing strategy:* Verbal effect-oriented rationalization

The strategies found by Burghausen and Balmer (2014) related to the heritage identity implementation have been employed in this study because, by identifying the specific manifestations of such strategies, it is possible to identify the preferred heritage identity that the organization wants to project. As the generation of empirical material has not been done through interviews and records but through juxtaposition of hypertext content, the implementation strategies of *performing* and *embodying* have been addressed only through their recurrent representations embedded in the *narrating* and *visualizing* strategies identified on the website.

As an initial step to this multimodal analysis, the data have been coded and transcribed in a Microsoft Word table document. It was necessary to make a separation between verbal and nonverbal resources in order to systematically look for patterns of multimodal representations that can appear at the level of each implementation strategy. Table 10.2 demonstrates the multimodal method of transcription and coding employed to record specific instances of representations of heritage identity implementation strategies.

5 Findings

According to the present analytical findings, it is evident that Coram succeeds to strategically use the resources of its historical repository for the representation of its identity in the present, "rather than in a purely retrospective sentimental way" (Burghausen and Balmer 2014: 2318). The central discourse that emerges from the representation of the four heritage identity implementation strategies on the website is that of evolving continuity. The multimodal construction of this discourse is sustained by a series of legitimations that are communicated not only by simply narrating and visualizing but also through the interplay of several modes: speech, written text, still images, moving images, sound, and music. These discursive means used by Coram to communicate about their heritage identity are rooted in complex multimodal patterns that generate "the conflation of past and present," "the conflation of old and new," and "the conflation of traditional and modern" (Burghausen and Balmer 2014: 2318). Table 10.3 showcases selected examples of heritage identity implementation strategies actualized in multimodal discursive form on Coram's website.

The following presentation of findings is divided in a discussion related to the verbal resources followed by the discussion of their interplay with nonverbal

Table 10.3 Selected examples of heritage identity implementation strategies actualized in multimodal discursive form on Coram's website

Heritage identity implementation strategies	**Multimodal representations of the heritage identity implementation strategies**		**Multimodal legitimating discursive strategies**
Performing	Milestone events (narrated and/or visualized)	Of the past	Verbal tradition authorization
			Still images
		Of the present	Verbal conformity authorization
			Still and moving images
	Practices, traditions and rituals (narrated and/or visualized)	Adherence to timeless values	Verbal tradition authorization
			Still and moving images
		Adherence to recurrent activities	Verbal conformity authorization
			Verbal effect-oriented rationalizations
			Black-and-white and color visualizations of similar activities from the past and present
Embodying	Famous supporters represented verbally and/or visually	The founding father	Verbal personal authority
			Painting of Thomas Coram
			Thomas Coram's playlist
		Perennial celebrities	Verbal expert and role model authority
			Painting of Handel
		Contemporary celebrities	Verbal expert and role model authority
			Photos of celebrities

(*Continued*)

Table 10.3 continued

Heritage identity implementation strategies	**Multimodal representations of the heritage identity implementation strategies**		**Multimodal legitimating discursive strategies**
	The "invisible" children, parents and workers represented verbally and/or visually	Of the past	Verbal effect-oriented rationalization Black-and-white images
		Of the present	Verbal goal-oriented rationalizations Color images
	Archives represented verbally and/or visually		Verbal impersonal authority Black-and-white and color images
	Buildings represented verbally and/or visually		Verbal tradition authorization Black-and-white and color imagesArchitecture blueprints drawings

resources in order to clarify the role of each resource for the multimodal representation of each heritage identity implementation strategy. Note that even though the findings are presented in this manner, this division entails a simplification of the analytical process in order to help introduce the method and the analytical findings in a systematic way. By first separating the semiotic modes and then integrating them in the following presentation, it is possible to offer more nuanced insights into their accumulated effects upon the strategies' multimodal representation.

5.1 Verbal resources

The dominant legitimations employed in the actualization of the heritage identity dimension are the personal, expert, and role model authority legitimations, and their specific occurrences on Coram's website are continuously marked by multimodal combinations. These legitimations are supposed to enhance simultaneously verbal and visual manifestations of the embodying and performing heritage identity implementation strategies. As already mentioned, this section is primarily concerned with their verbal manifestations. The effect of these authority legitimations is intensified as they are usually encountered in combination with the custom legitimations of tradition and conformity.

More specifically, in the representation of the embodying strategy, these legitimations are first related to the founding father, Thomas Coram, as well as to classic (Handel, Hogarth, and Dickens) and contemporary (Annie Lennox and Peter Capaldi) artistic celebrities who have supported and continue to support the organization.

Then, in connection with legitimating this embodying strategy of heritage identity implementation, similar past and present milestones in the organization's life, traditional activities, and rituals are legitimated. In this way, the performing strategy of heritage identity implementation is also represented and enhanced by this combination with the embodying strategy. For example, the most important milestone in the organization's past, namely Thomas Coram's initiatory campaign, is highlighted in the following example: "Thomas Coram began a campaign to create a home for these babies, overcoming widespread prejudice about children born outside of marriage, by enlisting the support of leading members of the aristocracy, the City, the arts and the sciences through a series of petitions" (Coram and the Foundlings Hospital 2016).

The continuity in adhering to the same timeless values in the organization's life through its founder's tireless campaign for children's welfare is also legitimated by mentioning that "Thomas Coram spent his life campaigning and he was still personally collecting signatures on petitions in his 80th year" (Thomas Coram asking for link 2016). These findings, related to the distinctive relevance of the founder for the articulation of the organization's heritage identity, are confirmed by other researchers who also note that founders' visions can become constitutive of an organization's identity and can remain present a long time after their death (Blombäck and Brunninge 2009; Hall 2004; Kimberly and Bouchikhi 1995). Furthermore, the visionary ideas of the founder link the past of the organization with a future unimaginable in those times: "Thomas Coram campaigned vigorously for education for girls—generally thought unnecessary and even dangerous at that time" (Thomas Coram on educating values 2016).

Another milestone is related to the organization's present, but it is still linked to the one mentioned above: "We also launched our charity's first online supporter pledge campaign to stand up for today's 'invisible' children, just as our founder Thomas Coram did 275 years ago" (Celebrate with us 2016). In both examples, through the tradition and custom legitimations, the time-bound words such as "began" and "first" suggest the passing of time and continuity. Renowned contemporary artists are also referred to and visualized in connection with the same supporting activities. For example, "Celebrities Annie Lennox and Peter Capaldi are among those pledging their support for Coram to stand up for today's invisible children" (Celebrate with us 2016). The legitimation of the continued devotion to children is strengthened by repeating the same metaphor, "invisible children," in relation to the long-standing tradition of supporting the organization.

Usually, evaluative words are also used recurrently in relation to the timeless values or time-bound milestones of the organization: "the pioneering spirit of the charity continues today," "a dazzling celebration of 275 years" (Celebrating 275 years of helping children 2016). The historical and continuous relevance of Coram's famous activities, for example the artistic therapies that help vulnerable children communicate, is evaluatively legitimated by referencing celebrities like Handel and Hogarth:

> We offer our ground-breaking art and music therapy at our Coram Community Campus in London, in schools and in the community, using music and art to stimulate self-expression in children, young people and their parents. These therapies are based on our rich musical and artistic heritage - the composer

> Handel and the artist Hogarth were two of Coram's earliest supporters. Find out more about the Hospital's famous supporters. (Our Creative Therapies 2016)

References to Handel and his legacy recur on the website in connection with other activities too: "Charity fundraising concerts are not a new venture for Coram. The composer George Frideric Handel launched the first charity concert for Coram in 1749" (Celebrating 275 years of helping children 2016).

The broad variety of verbal manifestations of both embodying and performing strategies of heritage identity implementation, which the organization employs, also includes personal authority legitimations found in famous classical authors such as Charles Dickens, who was also one of the organization's famous supporters and whose literary works have dealt with the fate of bereaved children. An excerpt from one of Dickens' most popular novels where he mentions Coram and his organization appears on the website: "The originator of the Institution for these poor foundlings having been a blessed creature of the name of Coram, we gave that name to Pet's little maid. At one time she was Tatty, and at one time she was Coram . . . and now she is always Tattycoram—Excerpt from *Little Dorrit*, Charles Dickens (Coram and the Foundling Hospital 2016).

The expert authority legitimations, on the other hand, are related to the lifetime commitment of those involved in the daily managing of the organization's activities. For example: "Jeanne Kaniuk OBE, Coram's Managing Director of Adoption, was awarded the prestigious silver Lifetime Achievement Award in the national Social Worker of the Year Awards 2014" (Our impact 2016). The three authority legitimation types may also be encountered in combination when stories of children helped by the organization are narrated. For example, John Brownlow who was provided a chance for a better life at the Foundling Hospital in his childhood subsequently dedicated his whole life to helping other children. He became a clerk at the Foundling Hospital and "rose through the ranks, becoming Treasurer's Clerk and later Foundling Hospital Secretary": "When he retired after 58 years of service, the governors praised his 'benevolent and charitable' life in which he 'devoted himself to the discharge of his duties with an energy and zeal beyond all praise and to the great advantage of the Hospital'" (The Foundlings Hospital 2016).

These personal, expert, and role model authority legitimations of continuous efforts, timeless values, and time-bound milestones of the organization are also accompanied by both goal-oriented (for example, "to stand up for today's 'invisible' children") and means-oriented (for example, "by enlisting the support") rationalizations. The consistency trait in the organization's activities is legitimated through effect-oriented rationalizations: "Since the charity was started 275 years ago, we've been helping children and young people develop their skills and emotional health, finding them permanent, loving homes and upholding their rights, creating a change that lasts a lifetime" (Our impact 2016).

Impersonal authority legitimations are also used with the verbal representation of the embodying strategy of heritage identity implementation. For example: "Historical records are available to the public through The Foundling Hospital Archives, held by the London Metropolitan Archives (LMA). With over 800 linear feet of shelving, records include the general registers, inspection books and petitions" (Researching Coram's archives 2016). Such impersonal authority legitimations that characterize the verbal representation of the embodying strategy (i.e., the archives) are meant not only to testify for the constant presence of the organization in the lives of vulnerable children but also to track down the constant development of its activities according to changing requirements. Therefore, implicitly, the performing strategy is also actualized through these impersonal authority legitimations: "Children's needs never change but understanding of the best ways to meet those needs has altered radically over the centuries. Our archives give a perspective on those changes, showing how Coram has reflected contemporary thinking and pioneered good practice in helping vulnerable children" (Our history 2016).

The embodying strategy of heritage identity implementation is also represented by combined references to even more prominent personalities (the King), laws (a Royal Charter), and buildings (the Foundling Hospital). Similar impersonal authority legitimations are employed: "Thomas Coram's 19-year campaign was finally brought to the attention of King George II who signed a Royal Charter on 17 October 1739 for the creation of the Foundling Hospital, which went on to be built in Bloomsbury, London, then surrounded by fields" (Our history 2016).

Verbal manifestations of the embodying strategy are also meant to reaffirm the links between past and present through combinations of both impersonal and personal authority legitimations. For example, "The Foundling Museum inspired a poetry anthology in 2012, Tokens for the Foundlings" (The Foundling Hospital 2016), in which vulnerable children, some of them quite famous today, commemorate the organization and its role in their lives. Such an anthology may also be interpreted as an overarching mythopoetic legitimation as it constitutes a symbolic embodying representation of the organization's heritage with a powerful performative dimension.

5.2 The integration of verbal resources and nonverbal resources

The performing and embodying strategies for implementing the heritage dimension of the organization are also legitimated on the website through other semiotic resources accompanying the texts. The accumulated effect of this interplay resides in the fact that it not only provides an immersion in the organization's past but also enhances a sense of continuity and immediacy. A visual enhancement of the discourse of evolving continuity is achieved through the paintings and photos of former and present famous supporters and through the succession of black-and-white and color close-up images of (adopted) children or young people in various stages of their lives. Portrait paintings and black-and-white and color photos of the famous former supporters accompany the legitimating texts. Furthermore, the multimodal legitimations also include Thomas Coram's playlist that gives the webpage users the possibility to listen to music. Music is also visually represented through images of Handel's manuscripts.

As far as the visual representations of vulnerable children or young people are concerned, even if they are isolated in close-up photos, the organization still legitimates its continuous impact through the photos' captions. For example, "More than one million children are helped by Coram every year" accompanies the close-up of a very young child. The repeated metaphor "invisible children" is visualized on the first page of the website through a half transparent black-and-white close-up of a child superimposed on the image of a brick wall. Apart from close-up images of children, the black-and-white and color photos of children together with their adoptive or divorced parents are also legitimating visualizations of the organization's continuous endeavors and successes. Furthermore, if the embodying strategy is represented through these portraits, the performing strategy is visually represented through black-and-white and color images of the children engaged in similar activities provided by the organization, both in the past and present. The black-and-white and color close-up images of children's hands playing various instruments visualize symbolically both embodying and performing strategies, as the heritage identity of the organization is manifested in the long chain of generations of children who have been given the chance to learn an instrument since Handel's time. The embodying strategy is also manifested through the combination of black-and-white images of the organization's buildings. Additionally, the immersion in the organization's past is facilitated not only through old images of the organization's buildings but also through the reproductions of the architecture blueprints that have been used.

The choice of replicating specific activities across time and, thus, strengthening the heritage identity of the organization is also visualized through moving images in the five videos that can be accessed on the above-mentioned pages. More videos are accessible at lower levels of subpages and on the organization's YouTube channel. The multimodal configuration of these types of dynamic texts provides the opportunity to intensify a sense of immediacy when viewing them, thus validating the continuous efforts of the organization and its core values. The flow of animated images, following or superimposed on drawings, images of the founder's statue, archive images, black-and-white and color images of children or young people, (adoptive) parents, and therapists, is accompanied by music and voice-over narration that highlights the exceptional vision and vows of the founder and the horrendous conditions from which the organization has saved generations of children. Each mode contributes to the multimodal blurring of time boundaries and consequently to the communication of the organization's perennial values that sustain its activities even today. The recorded, first-hand testimonials of present vulnerable children and young people, of therapists explaining and motivating their work, and of adoptive parents sharing their experiences, enhance the sense of immediacy and trustworthiness. The transcript for each film is also made available on the website in order to give the website's users the possibility to read the text if it has not been possible for them to understand the speech excerpts accompanying the moving images. In this way, the impact of the heritage identity implementation strategies is multimodally enhanced.

The complex affordances of the hypertext environment discursively strengthen each of the above-mentioned legitimation categories. These affordances facilitate the experience of overlapping and interweaving of layers of timeless or time-bound events through which the heritage identity of the company is kept alive. Coram's website is held together by a series of hypertextual means of communication, which include both label links and image links. Both types of links are supposed to contribute not only to the dynamic representation of the organization's heritage identity but also to the active involvement of the website's users. The users are encouraged not only to acquire more knowledge about the past, present, and future of the organization but also to actively support or/and engage in the wide range of present and future projects and recurrent activities of the organization.

The label links can identify a certain topic, and thus give the website's user the possibility to acquire more detailed knowledge about that specific topic and its heritage significance. Each of these topics is related to heritage identity implementation strategies and, through hyperlinking, the representation of these strategies is multimodally enriched because by following the links more multimodal knowledge is made available. On various pages of the website, these labels can identify topics related to the founder's timeless values that have always been sustained by the organization's activities ("Thomas Coram on educating girls"), topics related to ongoing services offered by the organization ("Child Law Advice Service"), or topics related to initiatory recurrent activities ("Our pioneering projects"). When the label links are verbalized through an imperative (for example, "discover," "visit our Coram Life education website to find out more," "learn more," etc.), the users are more strongly encouraged to profit from this knowledge enhancement possibility. The multimodal representation of both the embodying and the performing strategies of heritage identity implementation is reinforced through the presence of such links in the hypermodal environment of Coram's website. For example: "Follow the links on this page to find out more about our early supporters and the work of our inspiring Creative Therapies service including what types of creative therapy we offer, who can benefit from the sessions and who can make a referral" (Our creative therapies 2016).

In the example above, an enhanced conflation of the past, present, and future is realized when a combined representation of the embodying strategy ("our early supporters") and the performing strategy ("types of creative therapy we offer") is achieved through the provided cluster of links. Such examples confirm the findings of other researchers who also found that "as a communicative partner of the present, historical tradition belongs to the future" (Ericson 2006. 131).

Apart from these links, which encourage the users to acquire more knowledge, the organization provides links on various pages of the website, which encourage the users to get actively involved in or support various projects or activities: "come to an event," "fundraise for us," "donate now," etc. Their sense of belonging is enhanced when or if engaged in such activities because Coram, as other heritage organizations, "can harness positive public emotions" (Balmer 2013: 304) and provide a stable reference point in a changing world not only to the vulnerable children but also to all those involved. Encouraging the users to "be part of the story" on each page of the website, by sharing their interest in the context of various social media, the users also become involved in communicating the organization's heritage identity and in advocating for its timeless values and meaningful activities.

By using all these links, the collective memory of the organization is shared and passed on to more and more people. Although these links are obviously encouraging the users to involve themselves in present or future projects and activities, the information provided on the respective pages also makes specific references to the ways in which the respective projects and activities are either recurrent or have strong connections to traditional activities of the organization and its timeless values. In this way, once again, even if not directly, the heritage identity of the organization continues to be in focus.

6 Summary and discussion

The purpose of this chapter was to show how, by adopting a methodological synergy in a case study, it is possible to explore heritage identity implementation strategies represented in hypermodal context by a not-for-profit organization. Thus, a methodological process was laid out that can be replicated in order to approach the detailed analysis of such strategies by adopting a multimodal perspective. The present analysis was conducted by identifying the roles of specific semiotic modes and of their interplay in the representation of the strategies in a hypermodal context. The chapter provides a process view of how legitimations are employed in order to represent multimodally the strategies employed by a nonprofit organization for implementing its heritage identity.

The present multimodal analysis underscores how Coram is discursively constituted as a reliable historical organization by communicating its constant heritage identity implementation strategies hypermodally. After exploring the hypermodal data provided by Coram's website, it is evident that the embodiment of the organization's heritage identity manifests itself primarily via the personal identity of the founder and his everlasting values. Nevertheless, the other famous supporters, the successive generations of "invisible children," Coram employees, and various material things like archives and buildings also contribute to the nuanced representation of the embodying strategy. Simultaneously, the patterns of continuity and adaptation that surface from the multimodal representation and legitimation of performing strategies are combined through "overlapping and interwoven layers of timeless and time-bound events suggesting continuous progress" (Maier and Andersen 2015). The actualization of the heritage identity through these various events is represented on the website in order to illustrate the organization's lasting presence and reach.

The discursive coherence that sustains the representation of these heritage identity implementation strategies is preserved through the repetition of similar types of multimodal legitimations in various hypertextual combinations. These clusters of legitimations are supposed to reinforce the dominant discourse of evolving continuity across several semiotic modes and media.

When elaborating on the increasing prevalence of history in the corporate communication context, Suddaby, Foster, and Trank find that "history seems to hold a symbolic gravitas and legitimacy that other forms of persuasion simply cannot attain" (Suddaby, Foster, and Trank 2015: 163). As shown above, on Coram's website, the persuasive evocation of the organization's history receives new meanings when the heritage identity implementation strategies are communicated and legitimated multimodally. Past events from the organization's long history are repeatedly linked to the present and the future because the fundamental significance of the organization's long-term continuity is legitimated by connecting past, current, and/or recurrent events to timeless values and commitments. In this way, the strategically employed multimodal legitimations successfully communicate the trustworthy heritage identity of Coram and its enduring values and commitments. The representation of the heritage status of the organization is also built and reinforced hypermodally, as layers of legitimations come together through the links provided by the hypertext.

By identifying some of the key types of legitimations employed when the embodying and performing strategies are communicated, it is possible to delineate some of Coram's recurrent choices as far as the hypermodal representation of their heritage identity is concerned. However, as this is an exploratory study, a more nuanced understanding of heritage identity implementation strategies also necessitates interviews and other additional (archival) documentary materials for investigating the motivations behind selecting specific strategies. Furthermore, as this is a case study, the heritage identity implementation strategies of more not-for-profit organizations need to be explored comparatively in order to reveal possible multimodal patterns in the strategic communication of heritage identity, which are specific to this category of organizations. Therefore, the limitations of this study offer further directions for heritage identity research, as additional empirical scrutiny is needed for providing more methodological insights and maybe new theoretical articulations.

In this study, a methodological framework was proposed, combining multimodal discourse analysis with heritage identity theory that can be adopted by corporate communication scholars as an efficient tool for exploring the meaning-making potential of multimodal communication. Needless to say, the present application of this methodological framework is also a tentative demonstration that could encourage multimodality scholars to refine and extend their multimodal endeavors by combining them with methodological frameworks grounded in other theoretical domains.

References

Balmer, J. M. T. (2001), "Corporate Identity, Corporate Branding and Corporate Marketing – Seeing through the Fog," *European Journal of Marketing*, 35(3/4): 248–91.

Balmer, J. M. T. (2011), "Corporate Heritage Identities, Corporate Heritage Brands and the Multiple Heritage Identities of the British Monarchy," *European Journal of Marketing*, 45(9): 1380–98.

Balmer, J. M. T. (2013), "Corporate Heritage, Corporate Heritage Marketing, and Total Corporate Heritage Communications. What are They? What of Them?" *Corporate Communications: An International Journal*, 18(3): 290–326.

Balmer, J. M. T., S. A. Greyser and M. Urde (2006), " The Crown as a Corporate Brand: Insights from Monarchies," *Journal of Brand Management*, 14(1/2): 137–61.

Blombäck, A. and O. Brunninge (2013), "The Dual Opening to Brand Heritage in Family Businesses," *Corporate Communications: An International Journal*, 18(3): 327–46.

Blombäck, A. and O. Brunninge (2009), "Corporate Identity Manifested through Historical References," *Corporate Communications: An International Journal*, 14(4): 404–19.

Brunninge, O. (2009), "Using History in Organizations: How Managers Make Purposeful Reference to History in Strategy Processes," *Journal of Organizational Change Management*, 22(1): 8–26.

Burghausen, M. and J. M. T. Balmer (2014), "Corporate Heritage Management and the Multi-modal Implementation of a Corporate Heritage Identity," *Journal of Business Research*, 67(11): 2311–23.

Burghausen, M. and J. M. T. Balmer (2015), "Corporate Heritage Identity Stewardship: A Corporate Marketing Perspective," *European Journal of Marketing*, 49(1–2): 22–61.

Christensen, L. T. and G. Cheney (2000), "Self-Absorption and Self-Seduction in the Corporate Identity Game," in M. Schultz, M. T. Hatch, and M. H. Larsen (eds.), *The Expressive Organization: Linking Identity, Reputation and the Corporate Brand*, 246–70, New York: Oxford University Press.

Chreim, S. (2005), "The Continuity-Change Duality in Narrative Texts of Organizational Identity," *Journal of Management Studies*, 42(3): 567–93.

Coram and the Foundlings Hospital (2016). Available online: http://www.coram.org.uk/about-us/our-heritage-foundling-hospital (accessed August 30, 2016).

Ericson, M. (2006), "Exploring the Future, Exploiting the Past," *Journal of Management History*, 2(2): 121–36.

Gioia, D. A. (1998), "From Individual to Organizational Identity," in D. A. Whetten and P. C. Godfrey (eds.), *Identity in Organizations: Building Theory through Conversations*, 17–32, Thousand Oaks, CA: Sage.

Hakala, U., S. Lätti, and B. Sandberg (2011)," Operationalising Brand Heritage and Cultural Heritage," *Journal of Product & Brand Management*, 20(6): 447–56.

Hall, A. (2004), *Strategy Formation in the Family Business: In Search of Identity*, Paper presented at the 20th EGOS Colloquium, Ljubliana.

How we do it (2016). Available online: http://www.coram.org.uk/how-we-do-it (accessed August 30, 2016).

Kimberly, J. R. and H. Bouchikhi (1995), "The Dynamics of Organizational Development and Change: How the Past Shapes the Present and Constrains the Future," *Organization Science*, 6(1): 9–18.

Kress, G. (2010), *Multimodality. A Social Semiotic Approach to Contemporary Communication*, London: Routledge.

Kress, G. and T. van Leeuwen (2001), *Multimodal Discourse. The Modes and Media of Contemporary Communication*, London: Arnold.

Landow, G. P. (2006), *Hypertext 3.0. Critical Theory and New Media in an Era of Globalization*, Baltimore, MD: The John Hopkins University Press.

Lemke, J. L. (2002), "Travels in Hypermodality," *Visual Communication*, 1(3): 299–325.

Lemke, J. L. (2005), "Multimedia Genres and Traversals," *Folia Linguistica*, 39(1–2): 45–56.

Lemke, J. L. (2009), "Multimodality, Identity and Time," in C. Jewitt. (ed.), *The Routledge Handbook of Multimodal Analysis*, 140–51, London: Routledge.

Jones, R. (2012), *Discourse Analysis*, London: Routledge.

Jones, R. (2013), *Multimodal Discourse Analysis. The Encyclopedia of Applied Linguistics*, London: Blackwell.

Maier, C. D. and A. M. Andersen (2015), *The Strategic Communication of Corporate Heritage Identity in a Hypermodal Context.* Paper presented at The Conference on Corporate Communication, New York, US, June 2015.

Maier, C. D. and A. M. Andersen (2017), "Strategic Internal Communication of Corporate Heritage Identity in a Hypermodal Context," *Corporate Communications: An International Journal*, 22(1): 36–59.

O'Halloran, K. L. and V. Lim-Fei (2014), " Systemic Functional Multimodal Discourse Analysis", in S. Norris and C. D. Maier (eds.), *Interactions, Images and Texts: A Reader in Multimodality*, 137–55, Boston/Berlin: de Gruyter.

Our history (2016), Available online: http://www.coram.org.uk/about-us/our-heritage-foundling-hospital (accessed August 30, 2016).

Our impact (2016), Available online: http://www.coram.org.uk/difference-we-make/our-impact (accessed August 30, 2016).

Our creative therapies. (2016), Available online: http://www.coram.org.uk/how-we-do-it/our-creative-therapies (accessed August 30, 2016).

Royce, D. T. (2007), "Intersemiotic Complementarity: A Framework for Multimodal Discourse Analysis," in D. T. Royce and W. L. Bowcher (eds.), *New Directions in the Analysis of Multimodal Discourse*, 63–111, London: Lawrence Erlbaum Associates.

Suddaby, R., W. M. Foster, and C. Q. Trank (2015), "Rhetorical History as a Source of Competitive Advantage," *The Globalization of Strategy Research*, 27: 147–73.

The Foundling Hospital (2016), Available online: http://www.coram.org.uk/our-heritage-foundling-hospital/children-foundling-hospital (accessed August 30, 2016).

Urde, M., S. A. Greyser, and J. M. T. Balmer (2007), "Corporate Brands with a Heritage," *Journal of Brand Management*, 15(1): 4–9.

van Leeuwen, T. (2000), "Visual Racism," in M. Reisigl and R. Wodak (eds.), *The Semiotics of Racism: Approaches in Critical Discourse Analysis*, 330–50, Vienna: Passagen Verlag.

van Leeuwen, T. (2007), "Legitimation in Discourse and Communication," *Discourse & Communication*, 1(1): 91–112.

van Leeuwen, T. (2008), *Discourse and Practice. New Tools for Critical Discourse Analysis*, Oxford: Oxford University Press.

van Leeuwen, T. (2009a), "Critical Discourse Analysis," in J. Rekema (ed.), *Discourse, of course. An Overview of Research in Discourse Studies*, 277–92, Amsterdam: John Benjamins.

van Leeuwen, T. (2009b), "Discourse as the Recontextualization of Social Practice: A Guide," in R. Wodak and M. Meyer (eds.), *Methods of Critical Discourse Analysis*, 144–61, London: Sage.

Wiedmann, K.-P., N. Hennigs, S. Schmidt, and T. Wuestefeld (2013), "Brand Heritage and its Impact on Corporate Reputation: Corporate Roots as a Vision for the Future," *Corporate Reputation Review*, 16(3): 187–205.

11

The "Bologna Process" as a Territory of Knowledge: A Contextualization Analysis

Yannik Porsché

1 Introduction: What is contextualization analysis? Objects of investigation, questions, aims[1]

Fundamental terms of public debates are often used in a highly ambivalent way, and people's positions in these debates are not always clear. Participants constantly need to restate what exactly they are debating on and to who belongs to which camp. Nevertheless, researchers of social sciences frequently speak of a "media discourse," "immigration discourse," or a "Bologna discourse," as if the discourse existed independently of the wheres, whys, and whos of the discussion. In fact, one can identify different practices, depending on the institutional context or the epistemic culture[2], in which people negotiate what, for instance, a "Bologna discourse" is and who should participate in it. A discursive negotiation of this kind, thus, performs a positioning of the interactants, but it also goes beyond this, for example, when a conversation leads to a fixation of how to shape a European higher education area.[3]

For discourse research, this raises the question of how researchers could or should define a "Bologna discourse." The proposal of the microsociological approach, which lies at the heart of this text, is the following: neither can researchers do so, nor is the attempt worth pursuing. In this chapter, I present the analysis of contextualization in social interactions (referred to, in the following, in short as "contextualization analysis"). I do not suggest carrying out an academic debate that is analogous to ones found in everyday speech about how to define a "Bologna discourse" in distinction to other debates. This would entail a notion of discourse as a stable and homogeneous entity with clearly defined positions. In this view, discourses exist independently of how, when, and by whom they are articulated. This is not to say that such a notion of discourse does not constitute a helpful heuristic in everyday speech. Yet, in discourse research this would postulate the product, whose negotiation the research is set out to explore (see Sacks 1992: 42).

To specify what we are talking about seems useful in everyday speech (e.g., a debate about Bologna reforms), and it is helpful to situate people and institutions about and with which we talk in particular political, institutional, or other camps. However, if we choose to do so in research, this can give the impression that discourses or actors exist in a mythical macro world, in which they causally determine a European higher education area. In this case, a panel discussion on this topic would simply appear as a mechanical playback of previously set pawns in a game. At the same time, we cannot deny that on the micro level of everyday conversation we do refer to previously and elsewhere articulated discourses that we experience as enabling and constraining. This means we mobilize our knowledge about a topic of conversation and take into consideration what our conversation partners know about it and which perspective they take on it.

In order to take the above knowledge mobilization and consideration into account without having to posit the existence of a macro world, contextualization analysis micro-analytically deals with, for example, the role knowledge plays in social interactions. Contextualization analysis examines how participants of an interaction refer to knowledge about people, institutions, and discourses, and what functions these references and attributions have in and beyond the specific interaction. Everyday social interactions here appear not at all banal, since it is presupposed that they are the locus where societal relations of power are (re) produced. Seen from this perspective, power relations are not part of a context that is forged only in political voting or only through economic decision-making, but they are also fought out in institutionally framed social interactions.

The intertwining of specific interaction and context as well as discourse and materiality becomes clear in the analysis of diverse multimodal, that is, not only verbal but also embodied, contextualization cues. The microanalysis contains detailed descriptions of gaze, body movement, positioning in space, use of technology, and so on, and thus represents an approach to include all resources that participants draw on in public communication (Goodwin 2000, 2003; Haddington, Mondada, and Nevile 2013). In conversations, people frequently mobilize discourse developed previously and elsewhere. In order to bring absent communication "to life," participants employ contextualization cues. Contextualization also gives a certain materiality to what is said, for example, something can appear as an established fact or a spontaneous idea. At the same time, all contextualization cues are uttered in a particular way and at a specific place. For instance, spoken language implies particular intonation, or gesturing with an arm happens within certain bodily limits. Hence, we do not only experience discourses as enabling and restricting, but also we refer to discourses as well as people, activities, and institutions within material possibilities and limits. In multimodal media studies, this perspective attends to "media" in its most general sense of the word, that is, in terms of material, institutional, and other "conditions of possibility" (Ramming 2006: 9, 14). Technically mediated interaction includes the potential to reach a larger audience than the present audience (Broth 2008, Knorr-Cetina 2012). In this context-condition participants can address a particular public (Warner 2002)—in this way actively shaping the context—and refer to debates carried out elsewhere in the mass media. The analysis thus includes institutional interaction in the mass media (Clayman and Heritage 2002), but does not reduce the public sphere to these instances of communication (Porsché 2017). Instead, this method takes into account how participants attribute communication in the mass media the status of a more elevated public forum (Porsché 2016). Hence, contextualization analysis explores micro-analytically how one or more abstract macro levels are fabricated through the references of interaction participants. This way, participants in conversations generate positions that can be mobilized later on at other occasions (Scheffer 2007, 2014). Contextualization analysis thereby offers a heuristic to include the context in multimodal media studies with respect to the questions of which context to include and how.

2 Situating contextualization in the field of discourse analysis

The elements of contextualization analysis outlined in this chapter draw on ethnomethodological and post-structuralist instruments (Porsché 2012, 2013, 2016).[4] In the research literature, the former are mostly used for the analysis of interactions, and the latter for the analysis of discourses (cf. Wetherell and Edley 1999: 338). Due to occasionally divergent concepts of discourse and different conceptions of context's role in the analysis, these approaches are sometimes presented as incompatible. Some proponents from conversation-analytical (e.g., Peräkylä 2005: 881) or discourse analytical (e.g., Keller 2006) perspectives instruct researchers to keep to their specific expertise as conversation *or* discourse analysts. If at all, results should be brought together once the analyses have been carried out separately. I propose, instead, to switch between different perspectives already in the analysis. Only through changing between registers already during analysis does one pay enough attention to, and gain insight into, the entanglement of the micro and the macro. Furthermore, reflecting on this switching ensures that results are based on compatible theoretical premises. Contextualization analysis brings together elements from both research traditions in ethnomethodology and post-structuralism in order to analyze the "how" and "when" of contextualization, respectively. Conversation analysis, together with discursive psychology, offers a concept of discourse as an analytical basis for the empirical analysis of contextualization, subjectivation, and the representation of epistemic territories and institutions (Heritage and Raymond 2005; Wetherell 1998). The post-structuralist perspective clarifies how subjectivation, institutional structures, and positioning of people in epistemic territories regulate power relations (Angermuller 2014).

Ethnomethodological conversation analysis constitutes the first building block of contextualization analysis. The focus is on *how* (e.g., through spoken language, intonation, facial expressions, or gestures), in which sequential context, and with which function in the particular interaction activities are carried out. The contextualization analysis follows the established stance of conversation analysis that it is not up to the analyst to decide which meaning an utterance or an action has. Instead, the conduct of interaction participants is parsed for indices of their interpretation of previous utterance and their context, adhering to the principle of membership orientation (in contrast to, e.g., Bohnsack 2007; Oevermann et al. 1979). Ethnomethodology opens the perspective for and justifies the importance of standardized, situated practices and expectations that are "seen but unnoticed" (Garfinkel 1967: 36) in everyday life.

The analysis of (everyday) interaction constitutes the basis for the concept of discourse and context that has been developed in conversation analysis and subsequently in discursive psychology and contextualization research. Discourse is defined here as formal and informal spoken and written language—"naturally occurring talk and text" (Edwards and Potter 1992: 28). In multimodal analyses, it also refers to communication using the body, for example, gesture, facial expression, body posture, positioning, and movement in space (Goodwin 2000; Haddington, Mondada, and Nevile 2013). In this view discourse analytical research does not assume the existence of a plurality of countable and distinguishable discourses (Potter et al. 1990).[5] Instead, the analysis identifies smaller, situated "interpretive repertoires," that is, particular stylistic, grammatical arrangements that describe phenomena such as a person, an event, or an object. In contrast to discourses that researchers postulate, interpretive repertoires denote practices in which participants of conversations identify and understand descriptions as self-evident (Edley 2001). In a cultural community of communication, they designate a historical resource that one refers to through the use of metaphors, tropes, and figures of speech (Wetherell 1998). In the contextualization analysis here, I underline the differentiation between the perspectives of participants and researchers and reject discourse as an analytical category (in contrast to, e.g., Keller 2006; Knoblauch 2001; Parker 1990). At the same time, I take into account that interactants in everyday conduct assume that discourses circulate elsewhere in society prior to a particular conversation.

Approaches to research that build on ethnomethodology and that provide a foundation for microsociological contextualization analysis are institutional conversation analysis (Drew and Heritage 1992), positioning theory in social psychology (Harré et al. 2009), sociolinguistic contextualization analysis (Auer 1986; Gumperz 1982), and trans-sequential analysis (Scheffer 2010). These approaches make it clear that context is necessary for understanding an interaction and that context does not exist prior to, externally, or independently from a situation of interaction. I follow their proposal not to conceive of context as an entity that influences or even determines the interaction *a priori* as an independent variable (e.g., class, ethnic background, gender, institutional, geographic, or cultural setting). Instead, the term "contextualization" is used and conceptualized as an interactive and emergent process. This means that the context is only brought about and its effects unfold when interactants define a social situation, display their understanding to other participants by means of contextualization cues, and negotiate contradictory assessments and definitions of the social situation (Gumperz 1982: 131; Schegloff 1997; van Dijk 2006: 164–65). In the view of institutional conversation analysis, interactions are context-shaped and entail context-shaping activities. For instance, a conversation represents an academic panel discussion not because it takes place in a university but since participants carry out certain orders of speaking through their practices. In contrast to van Dijk's, Gumperz's, or Auer's approaches to contextualization research, in which mental models provide definitions of the context, microsociological contextualization analysis agrees with the criticism of cognitivism developed in discursive psychology (see Porsché and Macgilchrist

2014). This perspective rejects subjective participant constructs as explanatory models. Instead, the focus lies on public practices of social negotiation of representations (see Howarth 2006: 74) between interacting people and objects (Goffman 1974, 1981). This perspective conceptualizes utterances and activities as sequential context for the ensuing conversation. It allows articulating what constitutes expected conduct for a particular occasion of interaction.

In addition to ethnomethodological conversation analysis, post-structural or Foucauldian discourse analysis provides the second methodical building block for contextualization analysis. This perspective asks when utterances are possible, for instance, due to subjectivation and institutional configurations. The research focus is on power relations, created, for instance, through university disciplines, by defining who is allowed to speak with which legitimacy about what (Foucault 1980: 89). The focus on institutional configurations, in which utterances are said, regulated, and transformed complements the (very restricted) sequential understanding of context in conversation analysis with an understanding of the context as enabling and constraining conditions. Foucault's power-knowledge nexus is highly important for the contextualization analysis because in this line of research knowledge is not only understood in terms of rights and resources that have a repressive impact on subjects. Instead, subjects themselves are conceptualized as effects and mediums of power. Power here is understood as a relational, productive, and decentered network. It circulates and only exists in concrete micro-practices, which also enable and provoke resistance (Foucault 2003a: 397, 2003b: 236–43).

In light of Foucault's (2003a, b) notions that power is productively at work in micro-practices in all areas of life, that it is intertwined with knowledge, and that it is based on mechanisms of institutional exclusion, those ethnomethodological interaction analyses offer valuable empirical instruments, which focus on knowledge, institutions, and materiality. Sacks' (1979: 172) categorization analyses and Heritage's (2012a, b) analyses of epistemic authority provide a conversation-analytical foundation for the analysis of hierarchies and power relations. Sacks shows that categorizations imply power relations since categories can be "owned," enforced, or resisted. For example, adults typically refer to the category "youth" when they describe younger people. The described can, in return, invent their own self-descriptions. In Heritage's approach, the relative experience of differences in knowledge represents the decisive impetus for interactions. Identities are conceived here innovatively as attributions of knowledge (cf. problems and criticism of identity in discursive psychology in

Porsché and Macgilchrist 2014; see also Brubaker and Cooper 2000). Differences in knowledge can vary depending on "territories/domains of knowledge" (Kamio 1997; Labov and Fanshel 1977; Stivers and Rossano 2010) and moments of interaction. They are not based on an objective measure of knowledge but on the consensual expectation and assessment of the right that a particular person has a legitimate access to particular knowledge or information. In order to describe epistemic territories, Labov and Fanshel (1977) distinguish between "A-events" and "B-events," in which only person A or person B, respectively, has access to a piece of information. In a similar way, Pomerantz (1980) distinguishes between "type 1" and "type 2" knowledge, in which people in "type 1" can access from own experience, while "type 2" knowledge is based on "hear-say" or inference. Heritage (2012b: 7, 2012a: 32–33) introduces the distinction between "epistemic status" and "epistemic stance." The former denotes the more stable attribution of relative epistemic access to a domain of knowledge, and the latter refers to the more ephemeral positioning in a conversation (which can but need not concur with the former). Raymond and Heritage (2006) show, for instance, that other interactants define a grandmother as such by ascribing to her much knowledge about her grandchild (disregarding how much she actually knows about her grandchild). Together with studies in institutional conversation analysis, these instruments are used in the contextualization analysis for a comparison of epistemic cultures (Knorr-Cetina 2007). Combined with a Foucauldian perspective, an interaction driven by differences of knowledge appears as a political struggle that implies the possibility for resistance.

Multimodal interaction-analytic studies offer an additional methodical element to combine ethnomethodological instruments with a post-structural perspective that also pays attention to references to the material environment (e.g., Goodwin 1994: 626, Goodwin 2003). This establishes a basis for the contextualization analysis to include material dispositives in an empirical inquiry (Foucault 2000: 119–20). In contrast to other conversation analyses (e.g., Hindmarsh and Heath 2003) and in line with work by Goodwin (2003), I grant material things an important role in terms of them constituting an element in a co-participatory networking of humans and their material environment. With reference also to Latour (1994), contextualization analysis combines the contextualization of what is spoken *about* with contextualization practices *in the* interaction without assuming the existence of different analytical levels. Instead, the analysis explores the question of how participants construct different spaces and levels.

In sum, the ethnomethodological elements in the contextualization analysis conceptualize the interaction context as local and sequential unfolding in an orchestration by the participants. The embodied and material dimension of the context of interaction is oriented to multimodal approaches of analysis, some of which also take into account references to the material environment and its enabling and constraining impact. A perspective from discursive psychology and post-structuralism, however, relativizes the emphasis put on strategic and creative individuals in some conversation-analytical studies. In this view, subjects are only temporarily brought about in practices of subjectivation. Contextualization analysis' interest in the hierarchical contextualization of things that participants talk about (e.g., people, institutions, events, discourses) also goes beyond some ethnomethodological approaches by asking not only how reality is generated in interactions, but also what political meaning utterances have and which sociological consequences these constructions entail (cf. van Dijk 2006: 167). In its combination of ethnomethodological interaction analysis and post-structural discourse analysis, contextualization analysis thus explores the indexical production and relevance of the context dimensions of time, hierarchy, and materiality.

3 A sample analysis: A panel discussion on the "Bologna process"

The following example of contextualization analysis examines sequences from the final event in a series of panel discussions of the Colloquium Fundamentale. The colloquium, entitled "Education in the 21st century," took place in the summer semester 2004 at the Center for Applied Cultural Sciences and General Studies at the University of Karlsruhe.[6] The panel consisted of a moderator (in the following, "M"), an economy representative (Mr. Hünnekens, "H"), a principal of a private school (Ms. Ziegler, "Z"), and a representative of students (Ms. Scholz, "S"). The room also seated an audience, which is only occasionally heard and never visible in the recording. The panel discussion is transcribed word for word, including grammatical mistakes, hesitations, overlaps, unusual pronunciation, and so on. The level of detailing here is lower than in other varieties of multimodal conversation analysis (cf. Mondada 2008b).[7] A medium level of transcription that is similar to the one used in discursive psychology is only augmented by multimodal and prosodic details if the sequence appears to be of interest from a microsociological point of view, for example, if an explicit reference is made or if a code-switch, a misunderstanding, or a conflict cannot be identified in a less detailed level of analysis. This choice ensures that the analysis can also proceed to a consideration of what sociological consequences multimodal cues have. If the researcher were only to focus on the minuscule details of interaction, this would lead to highly interesting results from an interaction-linguistic perspective. However, this would not provide an analysis of the "bigger picture," such as how the interaction contributes to knowledge constructions and epistemic positioning. The notation used here can be found in Table 11.1. The multimodal signs show, first, what kind of description is referred to (e.g., direction of gaze or gestures), second, which person carries out the activity, and third, from when (and until when) the action is carried out in relation to the (spoken) language. For the analysis, I selected relatively short film sequences from the recordings of the panel that match the theoretical focus on subjectivation through the construction of epistemic territories. The following analysis aims to illustrate interaction-analytic instruments as a first step of a more encompassing microsociological contextualization analysis.

In the video recording of this panel discussion, several references are made to participants in the event, to the institutional frame of the interaction, to topics that had been raised in previous events, as well as to debates and perspectives that should be at the center of this debate. The following examples analyze the presentation of participants (example 1), in connection with the construction of epistemic territories (example 2) and different discursive levels (example 3). The contextualization analysis focuses on disagreements and generated consensus evident in practices of interaction on questions of *who* is speaking, *where* speech is carried out, and *what* is spoken about. We can call this process a hierarchical structured cartography of knowledge. Because these references aim to establish knowledge that all participants should take into account from this point on and because knowledge is intertwined with power, the analysis of these processes is always also an analysis of struggles of power relations.

The frame of the situation of interaction in which the construction processes take place is also generated—in addition to spoken language—through activities and sounds. For instance during the opening words of the moderator, coughing of one of the panelists and all sorts of preparatory activities that are visible in the video recording (e.g., sorting of papers, pouring water into glasses, straightening one's clothes) create the atmosphere of a panel discussion. Participants' conduct shows up norms that are associated with such a situation of interaction without the need to be postulated *a priori* in the analysis (e.g., the audience are relatively calm and only occasionally speak for a short time).

Table 11.1 Transcription notation for conversation analysis (adapted from Jefferson 2004 and multimodality in interaction adapted from Mondada 2008a)

(a) Conversation analysis

underline	emphasis
(.)	micropause
(0.4)	timed pause
[]	talk at the same time/overlapping talk
=	latching/next speaker continues with absence of a discernible gap
()	inaudible on the recording
((laughing))	described phenomena
Yea::h, I see::	extension of the preceding vowel sound
It's not right, not right AT ALL	words are uttered with added emphasis; words in capitals are uttered louder that surrounding talk
I think .hh I need more	a full stop before a word or sound indicates an audible intake of breath
[. . .]	square brackets indicate that some transcript has been deliberately omitted. Italics in square brackets is clarificatory information
>faster<	speaking faster than surrounding talk (>>much faster<<)

(b) Multimodality in interaction

Special sign beginning (& end) of gesture (e.g., gaze, hand)

here they refer to:

♦ to gestures and ■to direction of gaze of speaker M

✟ to gestures and □to direction of gaze of speaker Z

$ to gestures and € to direction of gaze of speaker H

small case initials added in multimodal notation line, if the movement corresponding to the speaker is not performed by the speaker him/herself but by a co-participant

---->	continuation of gesture in the following lines (with number to this line)
---->>	continuation of gesture until the end of the extract
>>	the gesture began before the beginning of the extract
# Fig. 1	screen capture at this point in time

3.1 Example 1: Construction of participants of interaction

In the following sequence at the beginning of the panel discussion (00:04:43.994–00:06:08.847)[8], the moderator introduces the discussants. An analysis of this sequence avoids speculations about who the participants of the interaction are and what their power relations are. Instead, the focus is on how the participants are positioned toward each other and toward the audience and the environment, including the camera and the institutional context. In other words, the analysis focuses on how subjectivations are performed in interaction and how, through an articulation of rights and expectations, the "starting position" or "frame" for the subsequent discussion is laid out. The analysis highlights those moments in which participants refer from the "here-and-now" of interaction to the past and to the future. First, it shows how present participants demonstrate knowledge or how it is ascribed to them through "imports." Second, it examines cues that characterize the knowledge.

Transcript 1

```
1    M: Ich darf
        I have the pleasure
2    Z: ((clears her throat))
3    M: dann (.) rechts von mir (.) .h Frau Doktor Hei- Professor, Doktor Heide Siegler
        now (.) on my right (.) .h to introduce Mrs Doctor Hei- Professor, Doctor Heide
4       vorstellen, in Sagen in Schlesion geborn. ■Ich ■ □werde □ >auch< ihr
        Siegler, born in Sagen in Silesia. I will >also< shorten your
                                                 ■to Z■
     Z:                                                □to M  □
5       Lebenslauf, ihr sehr interessanten Lebenslauf■(.) □etwas abkürzen □. Sie hat
        CV, your very interesting CV (.) somewhat. She studied
                                                   ■from Z to papers in hand-->8
     Z:                                                □ from M,laughs □
6       studiert alt [phil]ologie und Anglistik an der Universität (.) Göttingen,
        ancient philology and English linguistics at the University (.) of Göttingen,
7    H:             [((clears his throat))]
8    M: Promotion und Habilitation ■waren an der □Uni ■versität □ Wurzburg, sie hat
        PhD and habilitation were done at the University of Würzburg, she ehm was
                                   ■to Z              ■
     Z:                                         □ directed to M□
9       Gastprofessorinship äh gehabt an der Universität of Tax- Texas, aber auch in
        visiting professor at the University of Texas, but also at
10      Harvard (.) und ist (.) war von 84 Professorin für Amerikanistik und neuere
        Harvard (.) and is (.) was from 1984 professor of American Studies and
11      englische Literatur an der Universität. Und ■viele□hier im Raum ⇧werden auch
        contemporary English literature at that university. And many here in this room
                                                  ■to Z-->13
     Z:                                                □to M-->     ⇧arm to chin
```

```
12       sich dran ■erinnern und auch wissen, dass sie
         will also remember and also know, that she
                     ■ to papers -->
13    [...]
14       aber viele mehr werden (.) ⇧wissen, dass sie Gründungsrektorin ist von die
         but many more will (.) know that she is the founding director of the
      z:                            ⇧lowers arm, puts head on the side and folds hands-->
15       International University Bruchsal in Germany Bruchsal und eh ich denke das ist
         International University Bruchsal in Germany Bruchsal and ehm I think that is
16       ■ein Thema was hier an der Technische Hochschule sicherlich auch teilweise für
         a topic that here at the University of Technology surely also in parts caused
         ■ to Z-->18
17       energischen Streit auch (.) immer (.) wieder (.) mal  ⇧sorgt (.)⇧äh
         energetic dispute also (.) from (.) time (.) to time ehm
      z:                                                       ⇧nods      ⇧
18        Frau Siegler ■und das ist auch ein Grundprinzip ■hier von unserem ■Kolloquium
          Mrs Siegler and that is also a basic principle here of our colloquium
                       ■to papers                         ■to Z              ■to audience
19        Fundamentale möglichst unterschiedliche ■□Persönlichkeiten und ■Meinungen
          Fundamentally to bring together as different personalities and opinions as
                                                  ■to Z                  ■
      z:                                           □to M-->>
20        zusammenzubringen.
          possible.
```

In this first example (see Transcript 1), the moderator corrects herself twice in order to raise attention to the exceptional curriculum vitae of speaker Z (lines 3–5, called "self-repair" in Conversation Analysis [CA]). This way the fellow participant is granted authority by reference to previously and institutionally accredited merits. The moderator then calls upon knowledge supposedly shared by the panelists and the audience in the room ("many here in this room will also *remember* and also *know*," 11–12). Here, it is less important for the interaction analysis whether or not many participants in fact share this knowledge (whether the information is cognitively available and an act of remembering has been "triggered"). Instead, the analysis focuses on what function this reference to a supposedly shared knowledge has and how this knowledge is constructed. A shift of terms from "remembering" to a repeated "knowing" (12, 14) underlines that Z is not merely known due to an event (that one remembers), but that she is famous and thus part of a collective knowledge. These self-repairs and shifts of terms show that, in this context of interaction, it is considered necessary and appropriate to highlight Z's authority and that this is accomplished with reference to the "currency" of a common knowledge.

What is traditionally referred to as the "identity" of the participants can be formulated here as different claims and attributions of knowledge. By providing information and claiming that it is known to most of the audience, the moderator makes sure that those who in fact do not know Z or her current institutional position now do know why they could or should do so. This claim of supposedly shared knowledge also defines the audience, namely as one that is acquainted with the invoked past events and discussions. As with the question of which knowledge is *in fact* shared, the main question is not whether or not participants of the interaction *in fact* agree with a claim or not, that is, whether Z really expresses, for instance, an intrinsic will when she nods in agreement (17) or whether she is only behaving diplomatically. Instead of the question of authenticity, what is at stake when focusing on contextualization cues and reactions to them is reconstructing which epistemic territories are mapped in what way and with what functions and connotations. For instance, by nodding in agreement, Z affirms that the information provided about her and the context of interaction is correct. Through this affirmation and, among other things, through the order in which M directs her gaze to Z and subsequently Z turns to M (4, 8, 11), Z is not only positioned in the round of introductions. In addition, M is thus defined as a moderator, even without an introduction by name or a reference to her curriculum vitae. M therewith confirms that she did her research in order to correctly introduce Z, that it is her role to introduce the other people on the panel, and that she directs who speaks when to whom. The fact that Z takes on the role of affirming M's utterances also makes clear that Z, *qua* her "identity," is assigned privileged access to knowledge about herself (cf. Heritage 2011; Sacks 1984).

This example makes clear that M and Z contribute differently to their public positioning. This negates the natural assumption that each of the people present is responsible for his or her own positioning, or else that only the moderator performs the positioning. The interactive dynamic of subject positioning becomes visible, for instance, in the following sequences. Once that M and Z do not look at each other, Z laughs "to herself" in line 5. Arguably, she does not communicate with herself but rather displays to the audience that she is flustered. The influence of further participants becomes evident, among other things, in M saying "surely also in parts caused energetic dispute also (.) from (.) time (.) to time ehm" (16–17). The strong qualification "surely" in the claim of shared knowledge stands in contrast to the relativizing phrases "also in parts," "also from time to time" (16–17), and then the expressive particle "ehm" (17). By means of this juxtaposition, mitigating and hedging relativization, M contextualizes the statements so far as potentially problematic. Uttering the particle "ehm," she invites Z to agree and, thus, to contribute to the co-construction of painting a picture of which disputes occur at the university. Although theoretically speaking, Z would have the possibility to contradict M, agreement is projected as the preferred response, that is, it is suggested that agreement is sought (cf. Pomerantz 1984). From this perspective, none of the participants appear to be entirely autonomous in the practice of subjectivation. Instead, they contribute to positioning each other.

The construction of a shared knowledge about a dispute at the university enables the moderator, while introducing the panelists, to also define the university context and embed the colloquium in it (18–20). The connection between the specific interaction and the more general institution of the colloquium is performed, for instance, through M's direction of gaze: In line 18, M directs her gaze to her papers at that moment in which she refers to the basic principle of the colloquium. She does so after she directed her speech to Z by saying her name. It seems that a reference to the basic principle had already been noted before—and in this sense the principle appears to have been consequential before. As she focuses on the present interaction (saying "here," 18), M again directs her gaze to Z, before she looks toward the listening audience in the moment when she refers to the institution of the colloquium. In a similar way, M turns her gaze from Z to her papers while introducing Z (5–8). This characterizes the curriculum vitae as "sedimented" in paper, that is, as more stable and durable than the ephemeral situation of interaction. To a certain extent, this detaches Z's curriculum vitae further from the present person Z. Thereby, for instance, competences are ascribed to Z, disregarding whether or not she displays these in this particular interaction.

On the dimension of conversational sequentiality, these constructions of authority, situation, and knowledge constitute the sequential context for the following utterances. Furthermore, the reference to the PhD and professor titles represents hierarchical contextualization. Looking at the papers in which the "basic principle" (18) of the colloquium has apparently been materialized in writing and the reference to the interpretative repertoire of the curriculum vitae stabilize these phenomena as social facts. These contextualizations represent resources that are contextually given to a certain extent (e.g., gained degrees in the curriculum vitae or colloquium sessions that took place), yet these are always mobilized in a certain context of interaction, in which they serve a particular function. In this sense, they are only generated in the particular situation of interaction.

3.2 Example 2: Construction of epistemic territories

In the following example (00:19:27.975–00:22:09.321), the panelist H, who has also been introduced in the beginning (following the sequence of the first example), joins the ongoing discussion for the first time. He taps the moderator M on the arm while Z is speaking. M then interrupts speaker Z, once again introduces H, and announces his contribution. In connection with this practice of introduction, an epistemic territory is drawn where the participants are positioned as representatives of different camps.

Transcript 2

```
1   M: Darf ich  [ein ]fach
       May I simply
2   Z:           [Ja  ]
       Yes
3   M: an✝ diesem ■✝Punkt jetzt ■ >auch noch< Doktor Hönikens äh mit in die Diskussion
       at this point now >also as well< include Doctor Hönikens ehm into the discussion
    z:   ✝ nods,     ✝points to H
                   ■ to H        ■
4      einbinden und zwar ich äh hatte gesagt, er ist noch (.) in seine äh also äh
       namely I ehm said, he is still (.) in his capacity ehm well ehm
5      Tätigkeit jetzt bei der Kultur(.)stiftung der Allianz. Er wird (.) jetzt (.) ab
       now at the cultural (.) foundation of the Allianz. He will (.) now (.) from
6      (.) ersten Oktober sich auch vermehrt mit dieser Fragestellung aus einer völlig
       (.) the first of October have to, wants to, also deal more with this question
7      anderen Perspektive auch ähm kümmern müssen, kümmern wollen $, nämlich er wird
       from a completely different perspective, namely he will become
    h:                                                              $ nods
8      Rektor der >Akademie< (.) >der bildenden Künste< in Stuttgart und äh erstmalig
       principal of the >Academy< (.) >of Fine Arts< in Stuttgart and ehm for the first
9      eben auch ein Rektor der aus der >Wirtschaft< sozusagen, äh also, in diese
       time as I said also a principal from the >economic sector< so to say, ehm so,
10     Position >gewählt worden ist<. Herr Hönikens.
       >got voted< into this position. Mr. Hönikens.
11  H: $ Na, meine Wurzeln sind allerdings in der Kultur das haben Sie eingangs
       Well, my roots are, however, in culture, you
       $ crosses legs--->21
12     [ja auch gesagt ]
       also said that at the beginning
13  M: [das weiss ich  ]
       I know that
```

14 H: also ich bin ja auch, jetzt mit dem Hut der Allianz irgendwo ein
well I am indeed, now with the hat of the Allianz I somehow do

15 Wirtschaftsvertreter eh aber mit ner tiefen Verortung im Kultu:rraum
represent the economic sector ehm but with a deep positioning in cu:ltural space

16 ehm $ wir springen ja jetzt hin und her auch bei der $ Diskussion >und und< Sie
ehm as we are now jumping back and forth in the discussion >and and< you
$ points back and forth $

17 ehm weisen uns auch immer wieder eh darauf hin wo wir >aufhören sollen<. Ich
ehm remind us again and again ehm where we >should stop<. I

18 würde $ da ganz gerne mal kurz intervenieren ehm beziehungsweise Ihnen ehm
would quite like to intervene briefly here ehm or ehm would like to strongly ehm
$ points to Z

19 nachdrücklich ehm möchte Ihre Einschätzung nachdrücklich unterstreichen
underline your assessment

20 und $zwar jetzt $ tats #ächlich aus Sicht der$ Wirtschaft.
and now indeed from the perspective of the economic sector.
Fig. 11.1
$ raises finger,$ puts tie to side, $ sits back

21 ?: Mhm
Mhm

22 H: Eh die eh Unternehmen ✝ (.) in Deutschland #, eh das kann man ganz
Ehm the ehm businesses (.) in Germany, ehm one can very
Fig. 11.2, 11.3
holds knee, opens and closes elbows
Z: ✝puts hand to chin-->38

23 grob verallgemeinernd sagen, sind derzeit überhaupt noch nicht in der Lage eh
crudely generalise, are at the moment not at all able ehm

24 die ersten Abschlüsse Bachelor oder Master zu qualifizieren. Eh sie haben keine
to qualify the first Bachelor or Master degrees. Ehm they have no

25 Kriterien, um um die Qualität,$ wie sie selber sagen, auch tatsächlich richtig
criteria, to to actually assess the quality, as you say yourself, accurately.
$ points to Z

26 einzuschätzen.

27 [...]

28 H: MA sieht jawohl vor, also ich kann das jetzt auch nur referieren,
MA does plan exactly, well I can also only report this,

29 $ das wissen sie $ alle ja viel besser eh das man ähm äh berufsbegleitend ähm
all of you know much better ehm that one ehm ehm educates oneself ehm ehm
$ points to S & Z $

30 eh sich weiterbildet.
alongside a job.

31 H: Das ✝ setzt voraus, dass man, ja so hab ich's=
That presupposes, that one, yes that is the way I=
Z: ✝ puts head from one side to the other, raises arms-->>

32 Z: =Nicht unbedingt=
=Not necessarily=

33 H: =gehört, nicht?
=heard it, no?

Before this second extract from the panel discussion (see Transcript 2), the moderator commented critically on a reduction of the debate to the economic market. This can explain why H does not embrace him being positioned as "perspective of the economic sector" (20), without underlining his "positioning in cultural space" (15). Here, an image of a close and appreciative connection is created between the university and the cultural space, in contrast to the "outsider perspective" of the economic sector. These utterances make apparent which camps of argumentation are constructed and who is situated where and when in the interaction. In this connection, the contextualization analysis draws attention to the fact that neither the topography that is being brought about discursively nor the one that is reconstructed in the analysis represents an "objective" map of the discursive events.

In another context, one could imagine that H's economic position could be connoted positively. Here, however, several points suggest a problematization: a highlighting, a relativization, and a subsequent explanation of M ("from the >economic sector< so to say, ehm so, >got voted< into this position," 9/10) as well as a self-repair with respect to the constructed opposition between the economy and the cultural space or the university ("He will (.) now (.) [. . .] have to, wants to, also deal more with this question from a completely different perspective," 6/7). On the one hand, in this situation of conversation, it is necessary to highlight through intonation that someone in this circle represents the economic sector. On the other hand, it is appropriate to point out that he did not put himself in this different (but not necessarily bad) position.

Then H contradicts the positioning of M and highlights through an otherwise redundant utterance (seeing M's previous note) that he was previously active in the cultural space. His self-description of a "deep positioning in cultural space" (15) and the reference to "roots" (11) in culture stand figuratively in contrast to the economic positioning as merely a "hat" (14) that he put on. Only once H makes clear that he does not *only* want to be positioned as a representative of the economic sector, he continues his speech from an economic point of view. It seems appropriate to repeat what has already been said from this position ("as you say yourself," 25), that is, it is not only about what is said, but also who says it from which position and with which legitimation. The fact that H's position is generated by means of ascribed status of knowledge in contrast to other participants becomes clear in lines 28–33. Here, H notes "all of you know much better ehm that" (29). He points at S and Z when indicating that he knows less about this domain than they do. This also becomes manifest in Z questioning H's claim about MA degrees. A reference to "hearsay" knowledge (cf. type 2 knowledge in Pomerantz 1980) is mobilized to excuse an incorrect description of the Bologna reform. This furthermore implies that a representative of the economic sector does not need to know this—and, as a consequence of this, the economic sector is again located as distant from the university.

In the course of his positioning, H constructs the discussion as a "territory of interaction," in which he is positioned as an invited guest who is listening and keeping silent thus far. According to H, in this territory, panelists are jumping back and forth and risk being carried away from the "core" of the debate (16–17). In lines 20–26, H moves in his chair and, with this kind of code switching, initiates a new sequence of interaction (see Figures 11.1, 11.2, and 11.3 for illustration)[9]. Following introductory remarks with a raised finger (see Figure 11.1) and a gaze that was directed at M and Z, H puts his tie to the side (while he says "indeed," 20), leans back in his chair, grabs his knee, and in a "storytelling pose" turns to the audience. This contextualization is ratified, for instance, by Z who takes on a "listening pose" (hand to her chin, see Figures 11.2 and 11.3). The ratifying gestures indicate to the audience and to the camera that the other participants of the panel accept what H is saying and that it seems worthwhile listening to him (e.g., in contrast to instances in which they talk to each other simultaneously and do not focus their gaze on one person). Therewith positions in a wider discursive space between culture and the economy are also constructed on the level of moment-by-moment role taking of legitimate speaking and listening in a conversation.

In instances where utterances imply claims of truth and mechanisms of exclusion, an analysis of how long and with what legitimacy one is allowed to speak and in which circumstances speakers are heard constitutes an analysis of power relations (Spivak 1996, cited in Landry and MacLean 1996: 292). Moreover, in the panel discussion, several strategies appear through which participants attempt keeping or taking the floor as speakers. For instance, in sequence 01:26:39.260–01:27:55.854 participants negotiate how long people from the audience are allowed to speak in comparison to people on the panel

Figure 11.1 Speaker H raises finger.

Figure 11.2 Speaker H in "storytelling pose" and speaker Z in "listening pose".

Figure 11.3 Panelists listening to speaker H.

(e.g., one person from the audience continues talking even after the moderator interrupts him and says "the second important point was . . ."). Except for the situation in which M grants the opportunity to the audience to ask questions and comment, they are not able to participate in the discussion to the same extent as the panelists (also because it is more difficult for them to be heard without a microphone).

3.3 Example 3: Construction of discursive levels

In the last extract, the positioning of participants in epistemic territories of the economy and culture articulates which perspectives come together in the

"Bologna discourse." This focuses on the questions of who speaks for how long and who is listened to for how long. In the following sequence (see Transcript 3, 00:51:18.419–00:58:25.987), it also becomes clear how the Bologna process and higher education initiatives, that is, topics that are spoken about, are positioned hierarchically and materially in connection with the epistemic positioning of participants. The positioning is performed through pictographic locating of what is said on "levels" on which—in the view of the participants (in particular H in this example)—the discussion takes place. These levels are connoted in different ways. Since the context is structured through levels, power relations are negotiated in situ.

Transcript 3

```
1   H:  Ich glaube eine eine sehr europäische eh Idee, die natürlich (.)
        I think a a very European ehm idea, that naturally (.)
2       $>Bolo:gna< (.) auf ner $ Metaebene irgendwo auf(.)greift. Aber wir sind jetzt
        takes (.) up >Bolo:gna< (.) somewhere on a meta-level. But we are now
        $ hand higher level     $
        € >> to S-->3
3       $ nochmal€ zurück zu dem was wir $ bisher hatten € natürlich weit entfernt von
        again back to what we had so far of course far away from
        $ hand lower level                $
                  € to Z                                          € to S-->4
4       diesen € praktischen Diskussionen um hochschulpolitische Fragen. Wir
        these practical discussions about political higher education questions. We
                € to M
5       realisieren einfach mal europäische (.) Projekte.
        simply realise European (.) projects.
6   [...]
7       Sie haben recht und unrecht zugleich, wie das ja so oft im Leben ist.
        You are at the same time right and wrong, as it often is the case in life.
8   M:  ((laughs))
9   H:  Also in diesem Fall wissen Sie einfach zu wenig eh von diesem eh das ist kein
        Well in this case you simply know too little ehm about these ehm that is no
10      kein Strohfeuer und das ist kein kleines eh eh Pilotpro- projektchen, was dann
        flash in the pan and that is no small ehm ehm small pilot project, that will
11      wieder in den Schubladen verschwindet, sondern das hat längst eh auf
        then disappear again in the drawers, but that has already for long ehm
12      europäischer Ebene eine eh Qualität bekommen und ein ein eh .h und auch eine
        gained on a European level ehm a quality and a a ehm .h and also virtually a
13      benchmarke geradezu, die diesen Bolognaprozess begleitet, befördert und
        benchmark that accompanies, supports and intensifies this Bologna process.
14      intensiviert.
15  [...]
16      $ Ich kann,            $ kann nachvollziehen, dass sie schnell, auch ungeduldig
        I can, can understand, that you quickly, also impatiently
        $ patting hand to S   $
```

17 ähm ehm eh eine Ebene brauchen wo all das was sie an Fragen noch offen haben
need ehm ehm eh a level on which all those questions that remain open to you
18 schnell befriedigt wird, aber ein Campus Europae mit mit ehm eh in zehn Ländern
get satisfied quickly, but a Campus Europae with with ehm eh that has already
19 bereits realisiert zu haben [eh das] ist eine Dynamik,
been realised in ten countries ehm that is a dynamic,
20 S: [ja aber nur mit einzelnen]
yes but only with single
21 H: die sich [ja fortsetzen wird]
that will continue
22 S: [ja, aber das sind ja] wirklich nur eine Handvoll eh Hochschulen und
yes, but those are really only a handful ehm of universities and
23 wenn wir bis 2010 einen europäischen Hochschulraum realisieren wollen, .hh eh
if we want to realise a European Higher Education Area by 2010, .hh eh
24 das ist
that is
25 H: ähe
ehm
26 S: ehrlich gesagt doch nur ein Tropfen auf einen heißen Stein, wenn man mal so
to be honest only a drop in the ocean, if one wants to put it that way.
27 will. Ja also
Yes, well
28 M: Jetzt lassen wir Frau [Prof]essor Ziegler was dazu sagen
Now we let Mrs. Professor Ziegler say something to this point
29 Z: [Ja] .h ich verstehe den den Ansatz von Frau Scholz
Yes .h I understand the the approach of Mrs. Scholz
30 sehr gut, denn (.) ich meine, ich selber bin sehr dafür Eliten zu fördern und
very well, since (.) I mean, I am myself very much in favour of supporting
31 was Sie da tun ist eine Elite fördern.
elites and what you are doing there is supporting an elite.
32 […]
33 H: also Initiative, unsere Initiative ist ja ausdrücklich etwas Ergänzendes. Also
well the initiative, our initiative is explicitly something additional. So
34 alles [was] sie sagen, ist ja völlig richtig
all that you say, is entirely right
35 M: [ja, ja]
yes, yes
36 H: wir sind ja ganz bei einander.
we are in fact really close.

Participants here locate things they speak about on different "levels" between ideal "discourses" and material practices. It is in these practices that participants generate these levels (instead of the researcher postulating them). Moreover, people or institutions are positioned in relation to the levels generated. In lines 1–5, through his speech and gestures, H refers to a space that ranges from ideal and ironically uttered discussions about >Bologna< (2) to concrete questions of higher education and material practices that realize European projects (4/5). The constructed levels can be used for evaluative connotation: On the one hand, participants can portray something as a mere idea or an abstraction that is irrelevant for the non-academic world in contrast to down-to-earth realizations; on the other hand, something abstract can be presented as sophisticated and far reaching in contrast to mere application. The ironic contextualization in line 2 suggests the former interpretation, whereas the European level is presented positively and desirable in line 12.

H constructs the Bologna process as a large project worth supporting. It is, however, portrayed as still in its early days, and its abstract ideals are presented as something that can also amuse people. In several lines that are not reproduced here (6), S, in contrast, associates with the Bologna process hope for social justice that is still far from being realized. The contrasting and sometimes contradicting constructions of "Bologna" and its positioning on constructed levels are shaped by personal negotiations in interaction. The student representative S—in the section that is not shown in the transcript (6)—criticizes differences in prestige between institutions of education. H who represents the economic sector and the higher education initiative called "Cultural Foundation Allianz" understands this as a provocation and a criticism of his initiative. In a counterplea in which H portrays the European level as something positive (12), H from a parental and authoritarian perspective (see e.g., the patting hand movement, 16) states that the student S does not know enough about the topic (9). H attempts to add authority to his assessment by referring to a difference in age and knowledge. S resists this twice by interrupting H's talk (20, 22) without letting him speak again (see H's attempt in 25). Finally, Z is able to take on a mediating role between H and S and a consensus is constructed between the participants. As a "by-product" to the local negotiation of personal positioning of H, S, and Z and the power relations between them, the Bologna process is assessed and the initiative "Allianz" is ascribed a materiality of a "flash in the pan" (10) or a "drop in the ocean" (26) or else a lasting "benchmark" (13).

3.4 Results: Sociological relevance

The analyses of relatively short sequences above show how subjects are positioned in situations of conversation in terms of their epistemic status. This locates them in constellations of power in discursive territories. In connection to personal and institutional positioning, an epistemic territory is mapped with different levels of materiality (see Porsché 2017). Negotiations, such as this panel discussion about "Bologna," can potentially have an impact on higher education policy. Considering a longer time frame, the trans-sequential consequence of the panel discussion can be a shaping of educational politics and, thus, a modification of contextual conditions of an access to as well as the teaching and attribution of knowledge through education.

From a sociological point of view, the observation of an ambivalent use of the word "Bologna" is interesting with respect to the deficit in the EU's political legitimacy. Instead of searching for sweeping explanations of how people perceive and assess Bologna, it seems important to pay close attention to the situatedness and references to the context in interactions in which knowledge about Bologna is constructed. Looking at the functions that a reference to Bologna fulfills in a specific context of interaction (e.g., a counterplea in a conversation), it is clear that the "Bologna discourse" is not defined in and out of itself. Instead, its definition by and large depends on who articulates it in what moment of an interaction and in relation to which other constructions. It becomes clear that a detailed analysis of knowledge constructions cannot ignore the specific situatedness in the sequence of interaction. At the same time, the situation of interaction and the norms that are negotiated there can be understood as part of a more general institutionally created (political) "discussion culture." This is the case since the discussion only works through contextualization cues geared toward societal fields and discourses that—in the participants' view—exist elsewhere. These entities are produced locally and are connected with certain connotations. Particular speakers are furnished with authorities, roles, collective knowledge, and memories; "discourses" are invoked that the participants of a conversation did not make up; nevertheless, the contextualization analysis shows that these are not defined in a rigid and clear-cut way. Instead, participants articulate and negotiate them in local contexts of interaction. The analysis shows how participants try to fixate a discourse and people within it. At the same time the analysis points out practices of resistance against reducing people to certain positions (e.g., by H to be seen only as a representative of the economic sector or from S to be portrayed as an ignorant child). Similarly, attempts of fixating "discourses" about MA studies, an initiative in higher education, or the aim of Bologna, failed. Subject positions and power relations, thus, are neither ever set in stone nor can participants entirely abstract from what has been said before and elsewhere.

4 Further research procedure

On the basis of observations of how the Bologna discourse is characterized and invoked for which purposes at specific moments in conversation and in connection with ideas about the sociological relevance of these construction practices, the contextualization analyst generates further working theses and questions. These can, for instance, concern routinely constructed academic (as well as economic) subject positions or institutional-discursive levels and epistemic territories. These theses and questions can now be followed up by analyzing further sequences from this panel discussion and also in contrast to debates in other contexts (e.g., the press or the research literature). In an elaborate microsociological contextualization analysis, the examples would ideally be extended by further types of data (ethnographic field notes, interviews, documents, etc.) as well as additional analytical elements (e.g., an analysis of polyphony, cf. Angermuller 2014). This would allow incorporating further ways of contextualization and attending more to the atmospheres of discussions. Additionally, the panel discussion could be compared with later debates, for example, at the time in which the private university of the Founding Director Z was closed for financial reasons. This can be done in order to analyze the specificity and changes of institutional contexts. Building on insights about how construction processes work, sociological consequences, implications, and the interplay of constructions should be scrutinized both theoretically and empirically. The sequences of the example of analysis about "higher education discourse" could, for instance, generate the following research questions about ways and consequences of epistemic contextualization: How are debates carried out in other sequences or interactions? When do participants find a consensus in their understanding of the "Bologna discourse"? What significance do participants and researchers ascribe to the economic, academic, political, or institutional perspectives in other contexts? What role do different levels of abstraction of local and European higher education authorities play, for example, on a theoretical or practical economic level, or concerning the politics of higher education? Do participants refer to similar institutional positions, hierarchies, and personal involvements? What functions and impact does institutional positioning have in other construction processes of knowledge?

5 Conclusion: Possibilities, limitations, and further research

The analysis of epistemic contextualization shines a light on interactive dynamics with respect to how it (re)produces collective phenomena and power relations. Since the micro-analytic work assumes heterogeneity, variability, and dynamics, finding these might not be surprising. In contrast to approaches that instead postulate homogeneous and intersubjectively shared "discursive plates" or mental models, the antiessentialist concept of discourse used in contextualization analysis allows for a more solid empirical foundation. For media studies, this approach offers criteria and tools for including context in the analysis. In the multimodal analysis of case studies, contextualization can be identified and discussed in detail and with close connection to the empirical material.

In an approach that conceptualizes discourse as singular and not clearly delimitable, no claims can be made about different discourses. The distinction from approaches that adhere to a plural concept of discourse and that attempt to make claims about discourses using illustrative examples or statistical corpora thus constitutes the most obvious limitation of this approach. Instead of statistical representativeness that, for instance, lexicometric approaches can aim for, contextualization offers insight into "how-questions." In other words, lexicometric analysis can try to characterize the "Bologna discourse" in contrast to a reference corpus of comparable debates in the press. Multimodal contextualization analysis, in contrast, explores how Bologna is spoken about, pointed at, or otherwise oriented to in social interactions (or written about in texts), in order to understand how the interaction is managed and references are made to the specific context. The results of contextualization analysis are thus not tied or restricted to the topic of a "Bologna discourse." Instead, the results refer to the institutional context or the epistemic culture (Knorr-Cetina 2007) in which the debate is carried out. A multimodal approach attends to embodied resources as well as orientation in space and use of material objects such as audio and video recording for mass media broadcasting. Whereas a conservative multimodal conversation analysis that is at pains not to postulate what context is relevant restricts its focus to multimodal means of communication (e.g., to how someone points at something), the proposed contextualization analysis also attends to what participants point at. This does not mean the researcher needs to revert to speculations about what, for instance, a pointed-at Bologna process is. Instead, anchoring the analysis in contextualization cues enables the researcher to describe how participants characterize Bologna as, for example, something abstract or valuable and how they position people and institutions toward it.

In order to take this research perspective further, it is necessary that sociologists work together with sociolinguists, ethnographers, psychologists, and phoneticians with an analytical focus on interaction. In doing so, they can refine the work on multimodal contextualization cues and explore the relevance of their use for questions of politics and power. In this way, contextualization analysis can contribute in a significant way to the communication between multimodal conversation analysis and discourse analysis.

Notes

1 For very helpful comments on a German version of this chapter (Porsché 2014), I would like to thank Felicitas Macgilchrist and Martin Nonhoff.

2 Karin Knorr-Cetina defines epistemic cultures as "those sets of practices, arrangements and mechanisms bound together by necessity, affinity and historical coincidence which, in a given area of professional expertise, make up how we know what we know" (Knorr-Cetina 2007: 363).

3 The European higher education area was created in 2010 in the course of the Bologna process, which aims to standardize higher education in Europe. Students in all participating countries collect points according to a European Credit Transfer and Accumulation System in order to gain bachelor, master, or doctoral degrees.

4 Notes on the practice of the research procedure can be found in Porsché (2014).

5 However, contextualization analysis does consider every utterance to be plural in terms of its polyphonic and interdiscursive constitution.

6 The video recording is published by the organizers at http://digbib.ubka.uni-karlsruhe.de/diva/2004-362 (last accessed on September 19, 2016)

7 The presented analysis is constrained, first, due to limited time resources for transcription. Second, the low resolution of the film does not allow a detailed analysis of facial expressions beyond a rough description of the direction of gaze. For a fine-grained transcription, it is helpful to focus on different dimensions of interaction when repeatedly watching the sequence, for example, on spoken speech and body language of the primarily interacting and then including others. In the final version of the transcript only those observations are documented that are of use to the analysis and the presentation.

8 The video source can be accessed via the hyperlink in endnote vi.

9 In the first two frame shots of the panel discussion (Figures 1 and 2), only those people are shown whose gesture and body posture I focused on in the transcript and analysis in order to draw attention to these. Figure 3 shows the entire seating arrangement from which the silhouettes of speaker Z and H were cut out in Figure 2.

References

Angermuller, J. and J. Baxter (2014), *Poststructuralist Discourse Analysis. Subjectivity in Enunciative Pragmatics. Postdisciplinary Studies in Discourse*, Vol. 1. Basingstoke, Houndmills: Palgrave Macmillan.

Auer, P. (1986), "Kontextualisierung," *Studium Linguistik*, 19: 22–48.

Bohnsack, R. (2007), "Dokumentarische Methode und praxeologische Wissenssoziologie," in R. Schützeichel (ed.), *Handbuch Wissenssoziologie und Wissensforschung*, 180–90, Konstanz: UVK Verlagsgesellschaft.

Broth, M. (2008), "The studio interaction as a contextual resource for TV-production," *Journal of Pragmatics*, 40: 904–26.
Brubaker, R. and F. Cooper (2000), "Beyond 'Identity,'" *Theory and Society*, 29: 1–47.
Clayman, S. and J. Heritage (2002), *The News Interview: Journalists and Public Figures on the Air*, Cambridge: Cambridge University Press.
Drew, P. and J. Heritage (eds.) (1992), *Talk at Work: Interaction in Institutional Settings*, Cambridge: Cambridge University Press.
Edley, N. (2001), "Interpretative Repertoires, Ideological Dilemmas and Subject Positions," in M. Wetherell and S. Yates (eds.), *Discourse as Data: A Guide for Analysis*, 189–229, London: Sage.
Edwards, D. and J. Potter (1992), *Discursive Psychology*, London: Sage.
Foucault, M. (1980), "Two Lectures," in C. Gordon (ed.), *Power/Knowledge: Selected Interviews and Other Writings 1972-1977*, 78–108, Brighton: Harvester.
Foucault, M. (2000) [1978], *Dispositive der Macht. Über Sexualität, Wissen und Wahrheit*, Berlin: Merve.
Foucault, M. (2003a [1977]), "Das Spiel des Michel Foucault," in M. Foucault (ed.), *Schriften in vier Bänden. Dits et Ecrits. Band III 1976-1979*, 391–429, Frankfurt a.M.: Suhrkamp.
Foucault, M. (2003b [1977]), "Vorlesung vom 14. Januar 1976," in M. Foucault (ed.), *Schriften in vier Bänden. Dits et Ecrits. Band III 1976-1979*, 231–50, Frankfurt a.M.: Suhrkamp.
Garfinkel, H. (1967), *Studies in Ethnomethodology*, Cambridge: Polity.
Goffman, E. (1974), *Frame Analysis. An Essay on the Organisation of Experience*, Harmondsworth: Penguin.
Goffman, E. (1981), *Forms of Talk*, Philadelphia: Philadelphia University Press.
Goodwin, C. (1994), "Professional Vision," *American Anthropologist*, 96 (3): 606–33.
Goodwin, C. (2000), "Action and Embodiment within Situated Human Interaction," *Journal of Pragmatics*, 32: 1489–522.
Goodwin, C. (2003), "Pointing as Situated Practice," in S. Kita (ed.), *Pointing: Where Language, Culture and Cognition Meet*, 217–41 [draft version: 1–33], Mahwah, NJ: Lawrence Erlbaum.
Gumperz, J. J. (1982), *Discourse Strategies*, Cambridge: Cambridge University Press.
Haddington, P., L. Mondada, and M. Nevile (2013), *Interaction and Mobility. Language and the Body in Motion*, Berlin: de Gruyter.
Harré, R., F. M. Moghaddam, T.P. Cairnie, D. Rothbart, and S.R. Sabat (2009), "Recent Advances in Positioning Theory," *Theory & Psychology*, 19 (1): 5–31. DOI:10.1177/0959354308101417.
Heritage, J. (2011), "Territories of Knowledge, Territories of Experience: Empathic Moments in Interactio," in T. Stivers, L. Mondada, and J. Steensig (ed.), *The Morality of Knowledge in Conversation*, 159–83, Cambridge: Cambridge University Press.
Heritage, J. (2012a), "The Epistemic Engine: Sequence Organization and Territories of Knowledge," *Research on Language and Social Interaction*, 45 (1): 30–52.

Heritage, J. (2012b), "Epistemics in Action: Action Formation and Territories of Knowledge," *Research on Language and Social Interaction*, 45 (1): 1–29.

Heritage, J. and G. Raymond (2005), "The Terms of Agreement: Indexing Epistemic Authority and Subordination in Talk-in-Interaction," *Social Psychology Quarterly*, 68 (1): 15–38.

Hindmarsh, J. and C. Heath (2003), "Transcending the Object in Embodied Interaction," in J. Coupland and R. Gwyn (eds.), *Discourse, the Body, and Identity*, 43–69, New York: Palgrave Macmillan.

Howarth, C. (2006), "Social Representation is not a Quiet Thing: Exploring the Critical Potential of Social Representations Theory," *British Journal of Social Psychology*, 45: 65–86.

Jefferson, G. (2004), "Glossary of Transcript Symbols with an Introduction," in G. H. Lerner (ed.), *Conversation Analysis: Studies from the First Generation*, 13–31, Amsterdam/Philadelphia: John Benjamins.

Kamio, A. (1997), *Territory of Information*, Amsterdam: John Benjamins.

Keller, R. (2006), "Wissen oder Sprache? Für eine wissensanalytische Profilierung der Diskursforschung," in F. X. Eder (ed.), *Historische Diskursanalysen. Genealogie, Theorie, Anwendungen*, 51–69, Wiesbaden: VS Verlag.

Knoblauch, H. (2001), "Diskurs, Kommunikation und Wissenssoziologie,"in R. Keller, A. Hirseland, W. Schneider, and W. Viehöver (eds.), *Handbuch Sozialwissenschaftliche Diskursanalyse Band 1: Theorien und Methoden*, 207–24, Opladen: Leske & Budrich.

Knorr-Cetina, K. (2007), "Culture in Global Knowledge Societies: Knowledge Cultures and Epistemic Cultures," *Interdisciplinary Science Reviews*, 32(4): 361–75.

Knorr-Cetina, K. (2012), "Die Synthetische Situation," in R. Ayaß and C. Meyer (eds.), *Sozialität in Slow Motion*, 81–109. Wiesbaden: Springer VS.

Labov, W. and D. Fanshel (1977), *Therapeutic Discourse: Psychotherapy as Conversation*, New York, NY: Academic Press.

Landry, D. and G. MacLean (1996), *The Spivak Reader*, New York: Routledge.

Latour, B. (1994), "Une sociologie sans object? Remarques sur l'interobjectivité," *Sociologie du travail*, 36 (4): 587–607.

Mondada, L. (2008a), "Documenter l'articulation des ressources multimodales dans le temps: la transcription d'enregistrements vidéos d'interaction," in M. Bilger (ed.), *Données orales. Les Enjeux de la Transcription. Cahiers de l'Université de Perpignan. No 37*, 127–55, Perpignan: Presses Universitaires de Perpignan.

Mondada, L. (2008b), "Using Video for a Sequential and Multimodal Analysis of Social Interaction: Videotaping Institutional Phone Calls," *Forum Qualitative Sozialforschung*, 9 (3, Art. 39).

Oevermann, U., T. Allert, E. Konau, and J. Krambeck (1979), "Die Methodologie einer 'objektiven Hermeneutik' und ihre allgemeine forschungslogische Bedeutung in den Sozialwissenschaften," in H.-G. Soeffner (ed.), *Interpretative Verfahren in den Sozial- und Textwissenschaften*, 352–434, Stuttgart: Metzler.

Parker, I. (1990), "Real Things: Discourse, Context and Practice," *Philosophical Psychology*, 3 (2): 227–33.

Peräkylä, A. (2005), "Analyzing Talk and Text," in N. K. Denzin and Y.S. Lincoln (eds.), *The Sage Handbook of Qualitative Research*, 869–86, London: Sage.

Pomerantz, A. (1980), "Telling my Side: 'Limited Access' as a 'Fishing Device,'" *Sociological Inquiry*, 50: 186–98.

Pomerantz, A. (1984), "Agreeing and Disagreeing with Assessments: Some Features of Preferred/Dispreferred Turn Shapes," in J. M. Atkinson and J. Heritage (eds.), *Structures of Social Action: Studies in Conversation Analysis*, 57–101, Cambridge: Cambridge University Press.

Porsché, Y. (2012), "Public Representations of Immigrants in Museums. Towards a Microsociological Contextualisation Analysis," COLLeGIUM - Studies Across Disciplines in the Humanities and Social Sciences. Language, Space and Power: *Urban Entanglements*, 13: 45–72.

Porsché, Y. (2013), "Multimodale Marker in Museen," in E. Bonn, C. Knöppler, and M. Souza (eds.), *Was machen Marker? Logik, Materialität und Politik von Differenzierungsprozessen*, 113–51, Bielefeld: Transcript.

Porsché, Y. (2014), "Der 'Bologna Prozess als Wissensterritorium. Eine Kontextualisierungsanalyse," in M. Nonhoff, E. Herschinger, J. Angermuller, F. Macgilchrist, M. Reisigl, J. Wedl, D. Wrana, and A. Ziem (eds.), *Diskursforschung. Ein interdisziplinäres Handbuch. Vol. 2*, 379–403, Bielefeld: Transcript.

Porsché, Y. (2016), "Contextualising Culture: From Transcultural Theory to the Empirical Analysis of Participants' Practices," in J. Singh, A. Kantara, and D. Cserzö (eds.), *Downscaling Culture: Revisiting Intercultural Communication*, 311–36, Newcastle: Cambridge Scholars.

Porsché, Y. (2017), *Public Representations of Immigrants in Museums - Exhibition and Exposure in France and Germany*, Basingstoke: Palgrave Macmillan [in press].

Porsché, Y. and F. Macgilchrist (2014), "Diskursforschung in der Psychologie," in M. Nonhoff, E. Herschinger, J. Angermuller, F. Macgilchrist, M. Reisigl, J. Wedl, D. Wrana, and A. Ziem (eds.), *Diskursforschung. Ein interdisziplinäres Handbuch. Vol. 1*, 239–60, Bielefeld: Transcript.

Potter, J., M. Wetherell, R. Gill, and D. Edwards (1990), "Discourse: Noun, Verb or Social Practice?" *Philosophical Psychology*, 3 (2): 205–17.

Ramming, U. (2006), *Mit den Worten rechnen. Ansätze zu einem philosophischen Medienbegriff*, Bielefeld: Transcript.

Raymond, G. and J. Heritage (2006), "The Epistemics of Social Relations: Owning Grandchildren," *Language in Society*, 35: 677–705.

Sacks, H. (1979), "Hotrodder: A Revolutionary Category," in G. Psathas (ed.), *Everyday Language - Studies in Ethnomethodology*, 7–14, New York: Irvington.

Sacks, H. (1984), "On Doing 'Being Ordinary,'" in J.M. Atkinson and J. Heritage (eds.), *Structures of social action*, 413–29, Cambridge: Cambridge University Press.

Sacks, H. (1992), "Lecture 6 The MIR Membership Categorization Device," in G. Jefferson (ed.) *Lectures on Conversation, Vol. 1*, 40–8, Oxford: Blackwell.

Scheffer, T. (2007), "Event and Process. An Exercise in Analytical Ethnography," *Human Studies*, 30 (3): 167–97.

Scheffer, T. (2010), *Adversarial Case-Making. An Ethnography of the English Crown Court*, Amsterdam: Brill.

Scheffer, T. (2014), "Das Bohren der Bretter - Zur trans-sequentiellen Analyse des Politikbetriebs," in J. Adam and A. Vonderau (eds.) *Formationen des Politischen. Anthropologie politischer Felder*, 333–61. Bielefeld: Transcript.

Schegloff, E.A. (1997), "Whose Text? Whose Context?" *Discourse & Society*, 8 (2): 165–87.

Stivers, T. and F. Rossano (2010), "Mobilizing Response," *Research on Language and Social Interaction*, 43: 3–31.

van Dijk, T.A. (2006), "Discourse, Context and Cognition," *Discourse Studies*, 8 (1): 159–77.

Warner, M. (2002), "Publics and Counterpublics," *Public Culture*, 14 (1): 49–90.

Wetherell, M. and N. Edley (1999), "Negotiating Hegemonic Masculinity: Imaginary Positions and Psycho-Discursive Practices," *Feminism & Psychology*, 9 (3): 335–56.

Wetherell, M. (1998), "Positioning and Interpretative Repertoires: Conversation Analysis and Post-Structuralism in Dialogue," *Discourse and Society*, 9 (3): 387–412.

12

Afterword: Toward a New Discipline of Multimodality

Janina Wildfeuer and Ognyan Seizov

1 Multimodality in the (critical) spotlight

More than ever, multimodality is one of the most influential fields for analyzing media artifacts and human communication. It enjoys growing global popularity and a strong surge of interest from a number of disciplines in the humanities and beyond. Researchers and practitioners in linguistics, graphic design, visual communication, film theory, art history, conversational or interactional analysis, narrative studies, media studies, human-computer interface design, and many more are all finding the need to extend their areas of study beyond their original starting points in order to become *multimodal.* As already outlined in the introduction to this book, the number of academic publications in the field has grown dramatically over the last twenty years and demonstrates that multimodality is of contemporary theoretical, methodological, and analytical concern.

At first sight, it might seem as if the present volume adds to this as yet another edited collection of chapters related to multimodality in various ways and degrees. However, as it stands and as the individual contributions show in their own strength, the volume follows a different path, intended to diverge markedly from other edited collections and to be less mainstream and less rigid in the adherence to only one school or discipline. Instead, it covers perspectives from different traditions and new areas in multimodal research, which cross disciplinary and national borders. The selection reflects aspects of multimodal research that are relevant to the study of the twenty-first century's rich and complex media artifacts and those that will be evolving in the future. All approaches presented are on the cutting edge and able to tackle the challenges of old and new forms of communications existing and emerging today. With this, the explicit focus of the volume is one of readdressing, expanding, and applying central components of multimodality's genealogy. By reflecting on commonly held assumptions about the theory and practice of multimodal research in almost all chapters, the volume offers an in-depth critique of existing concepts across a wide range of different approaches to the study of multimodality—a venture that has not yet been undertaken with concerted intent in this still relatively young area.

In fact, multimodality's popularity does not imply universality or unconditional agreement: The conceptual anchoring of the notion, its theories and methodologies, as well as its empirical applications are not unproblematic, and the whole area brings with it certain frictions, disputes, and outstanding questions of principle. As Wildfeuer (2015) shows, for example, theoretical embeddings often remain nationally and regionally grounded, so that diverse and often ambiguous ideas coexist and develop individually. The differences between multimodal approaches in the Anglophone context in contrast to German-based discussions, for instance, are still surprisingly large, not least because of the concomitant inability to read and understand the respectively other language. The disciplinary (dis-)organization is surely another problem in the field of multimodality: Individual disciplines or orientations within disciplines address several different questions and ideas of the multimodal context by focusing on their very own media, for example, without fully concentrating on the overall paradigm, while still developing theoretically and methodologically valuable contributions. On the other hand, it must simply be stated that a discipline of multimodality is not yet fully developed to exist individually and independently alongside linguistics, communication studies, digital humanities, or other disciplines that are institutionalized via professorial chairs and study courses with the respective denomination. Multimodality, therefore, finds itself in the role of supplementing disciplinary and interdisciplinary work, which adds rather than replaces ways of dealing with the particular challenges and questions on all levels of description. At the same time, however, if the field is to move ahead, it has to reach a more mature status of reflection, mutual support, and interaction with regard to both past and future directions.

On the one hand, this critical approach to multimodality has definitely been advanced in this volume: All chapters strive to redevelop or provide anew theoretical concepts or empirical analyses, which are crucial to the study of multimodality from various perspectives and disciplines and with a view toward evolving issues of multimodal analysis. On the other hand, these individual approaches are only the beginning, albeit promising, of a deeper involvement with the field, which should then reinvent it into a well-grounded scientific discipline with significant implications for future research in all areas involved in the study of multimodality. We will outline in the following how this challenging endeavor can be taken further on the basis of the present volume.

2 Multimodality as a discipline

With its vital and broadening range of theoretical and practical issues addressed in the book, the volume indicates how the broad context of multimodality research is now on its way toward forming its own, fully acknowledged discipline, which demands and supports openness and motivation for largely inter- and transdisciplinary approaches and, at the same time, widens horizons and opens up fundamental new academic debates.

In fact, several individual approaches over the years and particularly more recently have already explicitly specified multimodality as a discipline without going into further details about, for example, the specific ways of thinking or scientific practices brought about by this particular discipline (cf., e.g., Constantinou 2005; Feng, Zhang, and O'Halloran 2013; Forceville 2013; van Leeuwen 2013, 2014). Their optimistic call, nevertheless, gives insight into "reliable methods" for the study of communication (van Leeuwen 2014: 282) as well as a potential to renew and rethink current ideas of media analysis (van Leeuwen 2013: 252). The critical aspect of confronting and discussing certain knowledge in the form of theories and methods as well as practical applications to the objects under study is, thus, a dominating trend to account for the disciplinary status of multimodality, which we now also pursue with this book.

In the Introduction to this volume, we have defined multimodality as a modus operandi for conducting research on human communication. We see this as a fundamental starting point for any further (critical) thoughts in the direction of a discipline of multimodality, which goes beyond previous definitions as, for example, "a theory, a perspective, a methodological application or a field of enquiry" (Jewitt 2009: 127) or "a field of application rather than a theory" (Jewitt 2014: 2). As a modus operandi, multimodality can now be generally understood as a habit of research, a method of operation (as the literal translation goes) that pushes the envelope and starts far-reaching discussions that cover description, terminology, methodology, as well as practical analysis in order to build the basis for a profound disciplinary status. As a consequence, we understand multimodality as an explicitly practical endeavor of engaging research and active exchange. It allows various disciplines and orientations as well as individual researchers to start thinking about their contribution to multimodal analysis by following the diversity and interconnectedness of theories, methods, and phenomena available and discussed over the last twenty-five years of multimodality research. Therefore, it explicitly invites us to broaden the knowledge and insights we already have on multimodality and its communicative practices and aims at bringing the multitude of approaches available in the field into the fold and to let disparate directions in theory and practice converge. The result will then be a common basis upon which the monolithic view of multimodality as a concerted disciplinary field can be built.

In order to develop its own strength, autonomy, and sovereignty as a new discipline, multimodality should then, as we wish, strive to characterize its area of research as follows (see also Krishnan 2009):

- in terms of the particular objects of research, that is, the multimodal phenomena addressed by and shared with other disciplines;
- with regard to its own accumulated knowledge of expertise about these phenomena and their analysis, which should, ideally, no longer be shared with other disciplines, but bring about new aspects of knowledge;
- on the basis of theories and concepts that specify their knowledge of expertise according to their own systematic organization;
- with regard to the specific vocabulary and terminology adjusted to both objects of research as well as theories and methods;
- on the basis of their specific research methods and requirements, which evolve from methods in other disciplines but develop their own specificities and freedom;
- and with regard to institutional manifestations in the form of study courses and professorial chairs at universities all over the world, academic departments, and professional and institutional associations.

To achieve these aims and to build the supporting pillars for this discipline, much work still has to be done! The multimodality context, however, with its broad range of theoretical and methodological approaches and its wealth of practical experiences is particularly well equipped to follow these paths and to establish itself as a strong and autonomous discipline, as we show exemplarily with this volume. The particular approaches presented call for theoretical evaluation and refinement as well as stronger developments and applications within larger, empirical projects. These aspects build important branches of the new discipline of multimodality, and we trust that this book paves the way for systematic and repeatable research into all kinds of multimodal phenomena from different perspectives and starting points. As example steps to implement the discipline, the chapters cross borders, expand views, and spark constructive academic debate. As a starting point, they build up a multimodal "micro-verse" whose constellations shape and form multimodality's disciplinary galaxy. The approaches, thus, show the general rich potential for undertaking further multimodal analysis, not only in the realm of a network-like structure with possible input from all research areas concerned with ways of communication, as we described it in Wildfeuer (2015), but also and in particular with regard to cogent disciplinary structures and a universal scientific rationality.

3 A further invitation to multimodality

We have shown in this volume that the present brand of multimodal research with its critical self-reflexivity and openness toward redefinitions and reconceptualizations is most apt to develop the needed strength and potential for discipline-building explorations. However, this endeavor is further dependent on the continuing motivation and effort of all researchers and analysts interested in multimodal phenomena who have until now relentlessly prepared the ground for solid multimodal work. In fact, we wish that the contributors to this volume were only a few among many other trendsetters who empower and bolster the discipline of multimodality and follow the targets and purposes proclaimed here.

Hence, as part of our continuous initiation of deeper inquiry into the study of multimodality, which we follow in publications as well as the organization of regular conferences at the University of Bremen, Germany (BreMM14, BreMM15, and—scheduled for September 2017—BreMM17), we again invite (following our first call in Wildfeuer 2015) all readers to not be hectored by our impetuous efforts but to enter the discussion, bring in their own perspectives, and develop and strengthen the manifold ways of approaching multimodal phenomena. As we have set out in the call for papers for the next Bremen Conference on Multimodality, we encourage every researcher interested in multimodal analysis to strive for unification and integration, toward the discipline of multimodality and to deal with the following example questions:

- What previously established disciplines should inform multimodality's disciplinary delineation? What is the place of, for example, semiotics, systemic functional linguistics, discourse analysis, interaction analysis, and other popular methods in the process of defining multimodality as a standalone discipline?
- Where can multimodality find its most inclusive and exhaustive theoretical basis? Do we need ways of combining the pioneers' work (see the introduction to this volume) to produce a new theoretical basis for the discipline? Do we start a new theory from scratch?
- What goes in multimodality's methodological toolbox? What existing empirical approaches define the field, how can we develop them further or combine them, and do we need new methods to capture multimodality's vastness?
- What are multimodal media and how do their various semiotic affordances shape multimodality within and across media formats? Are all media inherently multimodal?

Again, with regard to multimodality's general bandwidth, these questions are only starting points to begin both discussion and exploration of its disciplinary strength, and we welcome any initiative to develop this strength further. We think that only the vivid exchange of ideas and arguments for the discipline of multimodality, as exemplarily shown in this book, will allow its scope to grow further.

References

Constantinou, O. (2005), "Review Article: Multimodal Discourse Analysis: Media, modes and technologies," *Journal of Sociolinguistics*, 9(4): 602–18.

Feng, D., D. Zhang, and K.L. O'Halloran (2013), "Advances and Frontiers in Multimodal Discourse Analysis," *Contemporary Linguistics*, 1: 88–99.

Forceville, C. (2013), "Metaphor and Symbol: Searching for One's Identity Is Looking for A Home in Animation Film," *Review of Cognitive Linguistics, Multimodality and Cognitive Linguistics*, 19: 250–68. DOI:https://doi.org/10.1075/rcl.11.2.03for.

Jewitt, C. (2009), "An Introduction to Multimodality," in C. Jewitt (ed.), *The Routledge Handbook of Multimodal Analysis*, 14–39, London: Routledge.

Jewitt, C. (ed.) (2014), *The Routledge Handbook of Multimodal Analysis*, 2nd edition, London: Routledge.

Krishnan, A. (2009), "What Are Academic Disciplines? Some Observations on the Disciplinarity vs. Interdisciplinarity Debate," *Working Paper, University of Southampton, National Centre for Research Methods*. Available online: http://www.forschungsnetzwerk.at/downloadpub/what_are_academic_disciplines2009.pdf.

van Leeuwen, T. (2013), "Towards a Semiotics of Listening," in E. Djonov and S. Zhao (eds.), *Critical Multimodal Studies of Popular Discourse*, 251–64, London/New York: Routledge.

van Leeuwen, T. (2014), "Critical Discourse Analysis and Multimodality," in C. Hart and P. Cap (eds.), *Contemporary Critical Discourse Studies*, 281–96, London: Bloomsbury.

Wildfeuer, J. (2015), " Bridging the Gap between Here and There: Combining Multimodal Analysis from International Perspectives", in J. Wildfeuer (ed.), *Building Bridges for Multimodal Research: International Perspectives on Theories and Practices of Multimodal Analysis*, 13–34, Frankfurt a.M.: Peter Lang.

Author Index

Topic Index

www.ingramcontent.com/pod-product-compliance
Ingram Content Group UK Ltd.
Pitfield, Milton Keynes, MK11 3LW, UK
UKHW020243290926
15470UKWH00002B/719

9 781350 026513